THE COMING STORM:
Terrorists Using Cryptocurrency

By Steven Stalinsky

Preface: Lt.-Gen. (ret.) Vincent R. Stewart, former Deputy
Commander of the United States Cyber Command

ميمري
**MEMRI
BOOKS**

MEMRI Books
Washington, D.C.

The Coming Storm: Terrorists Using Cryptocurrency

By Steven Stalinsky

Published in the United States of America by MEMRI Books

www.memri.org | cjlab.memri.org

ISBN: 978-0-9678480-8-2 (paperback)
ISBN: 978-0-9678480-9-9 (e-book)
Library of Congress Control Number: 2019913923

ABOUT MEMRI

Exploring the Middle East and South Asia through their media, MEMRI bridges the language gap between the West and the Middle East and South Asia, providing timely translations of Arabic, Farsi, Urdu-Pashtu, Dari, and Turkish media, as well as original analysis of political, ideological, intellectual, social, cultural, and religious trends.

Founded in February 1998 to inform the debate over U.S. policy in the Middle East, MEMRI is an independent, nonpartisan, nonprofit, 501(c)3 organization. MEMRI's main office is located in Washington, DC, with branch offices in various world capitals. MEMRI research is translated into English, French, Polish, Japanese, Spanish and Hebrew.

ABOUT THE CYBER & JIHAD LAB (CJL)

The MEMRI Cyber and Jihad Lab monitors, tracks, translates, and researches jihadi and other types of hacktivist groups and activity, including secular, emanating from the Middle East, Iran, and South Asia, and studies jihadis on social media and online, with a focus on their use of encryption and other technologies. It works with tech companies to help come up with solutions for dealing with jihadis and terrorists online, and with legislatures to help develop laws for tackling this phenomenon. It also assists the business community in matters of cyberattacks and cyber threats from these sources.

cjlab.memri.org

TABLE OF CONTENTS

Preface By Lt-.Gen. (ret.) Vincent R. Stewart, Former Deputy Commander Of The United States Cyber Command And Special Advisor And Chairman Of The MEMRI Board Of Advisors

This study is the first significant research of its kind to show the scope of cryptocurrency use by terrorist organizations and their supporters, for fundraising and for financing attacks, purchasing equipment, supporting fighters and their families, and more. MEMRI research shows that over the past five years, this has developed from encouragement on jihadi blogs to donate "millions of dollars" in bitcoin to ISIS, Al-Qaeda, Hamas, the Muslim Brotherhood, and many other terrorist groups and their supporters soliciting donations in various cryptocurrencies on social media.

The report documents cryptocurrency use by a wide range of terror entities on many different online platforms. MEMRI research has found that the main platform for terrorist fundraising today is the encrypted messaging app Telegram – terrorists' "app of choice."

Terrorist groups regularly publish their bitcoin addresses when they solicit donations on online platforms –Telegram, Facebook, Twitter, and others – and also share detailed instructions via these platforms, in video, PDF, and other forms, to show potential donors how to donate in cryptocurrencies. This report reviews these, and also reviews the cryptocurrencies used by terrorists, which include, in addition to bitcoin, Dash, Ethereum, Monero, Verge, Zcash, and others.

MEMRI continues to be on the cutting edge of research on terrorist use of social media – notably on Telegram itself, and now with the looming threat of its imminent release of its TON blockchain and Gram cryptocurrency.

This landmark study shows undeniably that the future of terrorist funding and fundraising is happening now. The threat is too important to overlook; it must be researched and solutions must be arrived at, and this study is an important step in this direction. It will help educate those involved in the daily work of counterterrorism, and is a must-read for all those in government, the military, and academia who are concerned about this dangerous development.

* Lt.-Gen. (ret.) Vincent R. Stewart, former Deputy Commander of the United States Cyber Command, is Special Advisor and Chairman of the MEMRI Board of Advisors.

As ISIS's Caliphate has crumbled, and as it and other jihadi and terrorist organizations increasingly depend on the Internet, the most significant and dangerous recent development in global terrorism is their embrace of cryptocurrencies to fund their activities. This has quietly become part of their cyber arsenal, and to date there has been a lack of understanding of how to stop it.

The blockchain and cryptocurrency industry continues to gain momentum, and more and more companies allow the use of cryptocurrencies and even launch their own, it is understood, and expected, that criminal elements, including terrorist groups, will move even more quickly than they are today to use them. In the early days of social media, companies, particularly Twitter, YouTube, and Google, waited too long to address terrorist usage of their services, and authorities were two steps behind as well. This should not happen again, but in order to prevent it, it is vital that these companies put in place – now – clear policies banning criminal cryptocurrency activity on their platforms, with repercussions and penalties for companies that do not do so.

Such companies include, for example, Facebook, which on June 18, 2019 unveiled its Libra cryptocurrency, set for launch next year, which is backed by Visa, Mastercard, PayPal Holdings, and Uber Technologies; Samsung, whose new Galaxy phone includes a cryptocurrency wallet; Apple, which recently revealed its new software release with a "cryptographic" tool; Amazon, where purchases with cryptocurrency will soon be possible; and Airbnb, where rentals with cryptocurrency already are.

Facebook's announcement of its cryptocurrency was greeted with near hysteria. U.S. media have labeled it a "terrible idea" "untethered in ambition," and "a bold, bad move" and House Financial Services Committee chairwoman Maxine Waters said Facebook was "continuing its unchecked expansion and extending its reach into the lives of its users" – and her call for hearings on Libra was backed by the senior-most Republican on the committee, Rep. Patrick McHenry. Internationally, Bank of England Governor Mark Carney said Libra must be "subject to the highest standards of regulation"; French Finance Minister Bruno Le Maire encouraged finance leaders from G7 countries to prepare a report on it for next month; and German European Parliament member Markus Ferber said that Facebook could become a "shadow bank" and that regulators should be on high alert.

While concern about the threat of cryptocurrency use by bad actors, especially terrorist groups, is absolutely warranted, the focus should be not on Facebook but on the encrypted messaging app Telegram, known for some time to be a favorite of terrorists and jihadis, including ISIS, and already being used for fundraising by terrorist groups. This threat is now growing exponentially, as the platform is about to launch its own blockchain, TON, and cryptocurrency, Gram, reportedly by October 2019. Counterterrorism officials need to be prepared for the coming explosion of terrorist use of Telegram's cryptocurrency.

Although cryptocurrency fundraising by these terrorist and jihadi groups and their supporters continues to increase, many who are attempting to do something about it, including in government circles and academia, have not yet developed the tools to tackle it. Furthermore, recent research that has been widely cited by media has concluded that there is little indication that terrorist organizations are even using cryptocurrency extensively or systematically, and have downplayed any threat.

However, this report, "The Coming Storm: Terrorists Using Cryptocurrency," which focuses on jihadis' growing use of cryptocurrencies and the threat that this poses, documents how these groups and individuals are already extensively fundraising in cryptocurrencies – including for attacks and for weapons and equipment purchases, for food and necessities for fighters' families, and more, as well as promoting their use and discussing how to use them on various online platforms. It shows MEMRI's monitoring of their growing sophistication over the years in using them, and their promotion of them as being easy, safe, and secure to use.

The report presents op-eds on these groups' use of cryptocurrency that were published in *The Washington Post* and *The Hill* by MEMRI Executive Director Steven Stalinsky, on the threat posed by Telegram.

As the research for this report was being carried out, MEMRI staff have been briefing U.S. and Western government agencies,

media, academia, and legislatures about the ongoing findings. These efforts have also led to Congressional letters written with MEMRI's assistance, to Telegram CEO Pavel Durov insisting that he take immediate action against the well-known, well-documented widespread terrorist use of his platform by terrorists. The letters demanded, inter alia, that Durov "respond... with immediate urgency specifying a plan of action to combat extremist content on Telegram and create safeguards to prevent terrorist groups from using the platform as a secure fundraising tool."

This 200-plus page report comprises extensive and exclusive research, both published and not previously released, from the MEMRI Jihad and Terrorism Threat Monitor (JTTM) Project over the past two years. This research includes chatter by Western jihadis about using cryptocurrency, examines the use of cryptocurrency in fundraising campaigns and platforms, and also looks at how it is being promoted by jihadi and jihadi-linked organizations, tech groups, and individuals, from Al-Qaeda and the Islamic State (ISIS) to terrorist groups in the U.S. and the rest of the West as well as in the Middle East, Asia, and elsewhere.

Among the jihadi and terrorist groups examined in this report are pro-ISIS groups and the Al-Qaeda affiliate Hay'at Tahrir Al-Sham (HTS); the SadaqaCoins crowdfunding platform that links groups with donors using cryptocurrencies; the Al-Sadaqa organization which solicits donations in cryptocurrencies, and others. Hamas's military wing, the Izz Al-Din Al-Qassam Brigades, which is a U.S.-designated Foreign Terror Organization, is one of the terrorist groups most heavily promoting and using cryptocurrency, especially bitcoin, for fundraising, on platforms including Telegram, Twitter, and Instagram.

The report also includes discussions on online platforms and dissemination of tutorials on how and why to use cryptocurrency, as well as the sharing of bitcoin and other cryptocurrency addresses, including QR codes, for donating.

Many of these groups create dedicated social media accounts, on Telegram, Instagram, and other platforms, and hashtags for their campaigns, to disseminate attractive posters and graphics encouraging readers to donate, and listing prices of the equipment – ranging from uniforms to handguns to pickup trucks and RPGs – that their donations can buy. This report also details how prominent jihadi sheikhs, in the Middle East and the West, are promoting cryptocurrencies and depicting them as shari'a-compliant.

During the course of the research for this report, it was found that some groups, after receiving attention from media and authorities, apparently disappeared, along with their cryptocurrency wallets and addresses.

This could have been because they were shut down by authorities, because they switched platforms and changed their social media accounts, or because they moved deeper into the Dark Web.

This report is accompanied by a short video, "The Coming Storm: Terrorists Using Cryptocurrency - Video Compilation," that can be viewed on the MEMRI Vimeo page (https://vimeo.com/memri/the-coming-storm-terrorists-using-cryptocurrency). It shows research from the report on jihadi groups and individuals actively involved in fundraising in cryptocurrency.

About this report:

The following report presents examples from content shared by the many jihadi groups that are actively using cryptocurrencies, in fundraising, for weapons and equipment purchases, and for many more purposes. It includes research by the MEMRI JTTM team and editors over the past two years, among them R. Sosnow, M. Khayat, M. Al-Hadj, R. Green, S. Benjamin, N. Mozes, A. Smith, J. Goldberg, and C. Caruso.

In a December 17, 2018 op-ed in *The Washington Post* titled "The Cryptocurrency-Terrorism Connection Is Too Big To Ignore," MEMRI Executive Director Steven Stalinsky wrote:

On Nov. 26 in a federal court in New York, 27-year-old Zoobia Shahnaz pleaded guilty to financially supporting the Islamic State terrorist group with a scheme that employed money laundering and bank fraud, along with bitcoin and other cryptocurrencies, according to prosecutors. She was tripped up when law enforcement officials detected overseas wire transfers designed to avoid financial-reporting requirements.

Cryptocurrency has come to terrorism, with an array of terrorist organizations exploiting the anonymity afforded by blockchain technology for fundraising and finances, yet U.S. counterterrorism officials appear to have been slow to grasp the extent the problem.

Certainly the 9/11 Commission Report in 2004 recognized that "vigorous efforts to track terrorist financing must remain front and center in U.S. counterterrorism efforts," in part because "information about terrorist money helps us to understand their networks, search them out, and disrupt their operations." But that was long before cryptocurrency emerged as a method for moving money while evading detection. The National Strategy for Counterterrorism released by the White House in October — the first such report since 2011 — could have been expected to address the role of cryptocurrency in terrorist financing, but it didn't.

The new national strategy noted that terrorists employ encrypted communications, and it vowed to "deny terrorists the ability to raise funds" and "plot attacks, travel, and abuse the global financial system." But in a gaping omission, the national strategy failed to link encryption and terrorist funding.

With the Islamic State's physical caliphate in shambles, revenue from oil and taxes have disappeared, but cryptocurrencies such as bitcoin, Dash, Ethereum, Monero, Verge and Zcash, with others in development, constitute an alternative funding source for the terrorists. Transactions are swift and anonymous, and disrupting them is difficult. In addition to more-established terrorist organizations, an emerging cadre of terrorist groups and their affiliates, such as Al-Sadaqah, Malhama Tactical and the Ibn Taymiyyah Media Center, have begun using cryptocurrency. Communications about transactions

The Coming Storm: Terrorists Using Cryptocurrency

often take place on encrypted messaging apps, such as Telegram, favored by terrorist groups because they are easy to use and offer a secure venue for planning and recruiting — and for advising Western supporters about how to use cryptocurrency.

Telegram's encrypted text and voice messaging have gained worldwide popularity since its launch in 2013 by Russian entrepreneur Pavel Durov. The service passed 100 million active monthly users two years ago, with its secrecy also inevitably attracting criminal and terrorist organizations. By far, these groups rely on bitcoin for financial activity on Telegram, but now Durov appears set to launch Telegram's own cryptocurrency after obtaining $1.8 billion in funding.

The prospect has alarmed Rep. Ted Poe (R-Tex.), chairman of the House subcommittee on terrorism, nonproliferation and trade, and Rep. Brad Sherman (D-Calif.), ranking member of the subcommittee on Asia and the Pacific. On Oct. 25, they wrote to Durov expressing concern that the launch of a Telegram cryptocurrency "will make it even easier for terrorists to fundraise without disruption." The congressmen asked Durov to provide a "plan of action" to "create safeguards to prevent terrorist groups from using the platform as a secure fundraising tool." Poe's office says Durov hasn't replied.

Other alarms about terrorists' use of cryptocurrency have been sounded in recent months. In September, the House approved a bill introduced by Rep. Ted Budd (R-N.C.) calling for the establishment of an independent financial technology task force to research terrorists' use of new financial technologies and specifically to combat the use of cryptocurrency by terrorists. (The bill awaits Senate action.) The intergovernmental Financial Action Task Force, which combats money laundering and terrorism financing, noted in an Oct. 19 statement that the Islamic State and al-Qaeda are using cryptocurrencies and called on governments worldwide to establish rules for their use.

More leadership from those working worldwide against terrorist fundraising is urgently needed. It should not take a major terrorist attack, planned on an encrypted apps and financed with cryptocurrency, to get their attention.

On March 8, 2018, MEMRI Executive Director Steven Stalinsky published an op-ed in *The Hill* titled "Terrorists Have Been Using Bitcoin For Four Years, So What's The Surprise?" The following is an expanded version of the op-ed, which includes notable developments since the original publication.

As the value of Bitcoin has reached new highs, there has been a major increase in attention to it, including news articles and TV specials – along with concern from governments worldwide that terrorist groups are using digital wallets to send and receive funds in Bitcoin, the decentralized peer-to-peer cryptocurrency created in 2009. With ISIS's physical caliphate crumbling along with its own once-heralded economy made up of its own currency and banks, utilizing Bitcoin as a backup financial plan is now in play.

In recent months, French, Italian, and U.K. legislators have sought more regulations on Bitcoin because of potential terrorist usage. In the U.S., there have been bipartisan Congressional hearings and legislation. In May, U.S. Rep. Kathleen Rice (D-N.Y.) said, when she introduced H.R.2433, the Homeland Security Assessment of Terrorists' Use of Virtual Currencies Act, that "with groups like ISIS becoming more technologically sophisticated

The Coming Storm: Terrorists Using Cryptocurrency

STUXNST ☑
@seaunx

Soon SEANUX v1.0 by @Official_SEA16
suport from dominican republic. For
donations
1D1PkRGVC5ATBMafKgM2PEwHHmsAhQ
vDDy

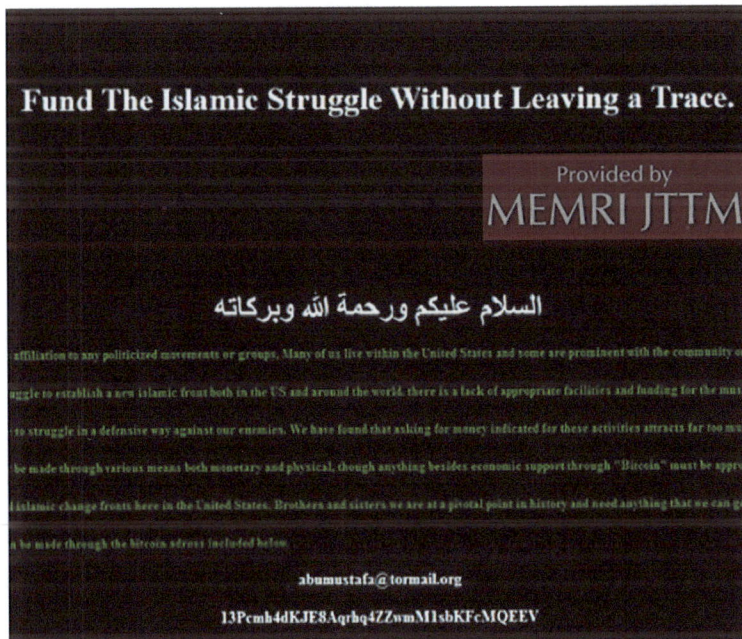

Fund The Islamic Struggle Without Leaving a Trace.

السلام عليكم ورحمة الله وبركاته

affiliation to any politicized movements or groups. Many of us live within the United States and some are prominent with the community on

uggle to establish a new islamic front both in the US and around the world, there is a lack of appropriate facilities and funding for the musli

to struggle in a defensive way against our enemies. We have found that asking for money indicated for these activities attracts far too much

be made through various means both monetary and physical, though anything besides economic support through "Bitcoin" must be approve

islamic change fronts here in the United States. Brothers and sisters we are at a pivotal point in history and need anything that we can get

be made through the bitcoin adress included below.

abumustafa@tormail.org

13Pcmh4dKJE8Aqrhq4ZZwmM1sbKFcMQEEV

and virtual currencies becoming more accessible, the table is set for this threat to grow significantly... We need to confront this threat immediately." In January, Rep. Ted Budd (R-N.C.) introduced H.R.4752, the Financial Technology Innovation and Defense Act, to establish a task force combat terrorist use of cryptocurrency.

While these are positive developments, terrorist groups were experimenting with Bitcoin as early as 2014. One of the earliest examples was that year by the notorious Syrian Electronic Army which retweeted a user from the Dominican Republic soliciting donations in Bitcoin in order to raise money to pay for helping to distribute the organization's self-designed Linux software.[1]

Also in 2014, a Dark Web site, "Fund The Islamic Struggle Without Leaving a Trace," was encouraging supporters to donate in Bitcoin to fight the U.S. The one-page website contained a message in English and a Bitcoin address to which users could send donations, and also provided an individual's address for TorMail, which was an anonymous webmail service that operated in the Dark Web until the FBI seized its service database following the Freedom Hosting Incident.[2]

Other early cases of terrorist groups' use of Bitcoin include: in 2015, reports that the Charlie Hebdo attacks in Paris were funded in Bitcoin by al Qaeda in the Arabian Peninsula (AQAP), with anti-terrorism hacktivist Ghost Security Group having uncovered a digital wallet containing $3 million in Bitcoin value believed to be financing the attacks for ISIS;[3] in 2016, reports that Bitcoin, including a wallet containing $3 million in Bitcoin, could have been used in the November Paris attacks that year.

Also, "Fund the Islamic Struggle Without Leaving a Trace," a pro-ISIS forum, began posting links in June 2015 to a Dark Web page that provides a Bitcoin address through which funds can apparently be sent to ISIS.

Some of the ISIS supporters in the West using Bitcoin who thought they were shielded by its anonymity have been apprehended. In 2015, 17-year-old Virginian Ali Shukri Amin pleaded guilty to using Twitter to teach ISIS members how to use Bitcoin. On July 7, 2014, Amin (@AmreekiWitness) tweeted a link to an article he wrote and posted on his blog, titled "Bitcoin and the Charity of Jihad," explaining how Bitcoin works and how to use the anonymity-providing "Dark Wallet." Muslims, he wrote, should use Bitcoin to support jihad because it has "no points of weakness" and is untraceable by "kafir government."[4]

More recently, in December 2017, 27-year-old Long Island resident Zoobia Shahnaz was charged with money laundering and bank fraud; authorities say she stole and transferred around $85,000 to support ISIS, with some $62,700 of it in Bitcoin and other cryptocurrencies.[5]

Highlighting jihadis' growing interest in Bit-

Provided by
MEMRI JTTM

The Coming Storm: Terrorists Using Cryptocurrency

coin, last month the pro-Al-Qaeda English-language magazine Al-Haqiqa devoted an entire article to using Bitcoin, and examined the shari'a permissibility of using it and other cryptocurrencies to fund jihad. It noted: "We see lots of potential for the use of cryptocurrencies for our purposes."[6]

Also highlighting how terrorist groups are using Bitcoin is the case of the Isdarat website, a hub for ISIS content; on November 30, Telegram users linked to it and explained how to donate to it in Bitcoin. Two weeks later, a reader replied that the funds donated in this way had bought computers for jihadis.[7]

The most important organization to openly use Bitcoin is the Syria-based anti-Assad Al-Sadaqah organization, which promotes itself as "an independent charity organization that is benefiting and providing the Mujahidin in Syria with weapons, financial aid, and other projects relating to Jihad." Its ongoing fundraising campaign in Bitcoin on social media is aimed at Americans and other Westerners – and a growing number of them are donating, disseminating its posts, and chatting about how to donate. The campaign shares its Bitcoin wallet number on its Telegram channel, which it opened on November 8. A recent appeal noted, "If anyone has a Bitcoin ATM in your area of country, then you can send money 100% anonymously."

Additionally, we have found one pro-ISIS tech group recommending the use of Zcash over Bitcoin because, it said, Zcash was harder to trace. We have also identified a new Telegram channel soliciting funds in bitcoin for supporting jihadis' families.

Post on Telegram: Bitcoin "does not provide a strong level of privacy"

Ummah Reports
Forwarded from Sabranyanafsi

السلام عليكم ورحمة الله وبركاته

hope this reaches you in the best of health and iman. I am a brother here in sham. If you know anyone that wants to help support families here fisaballah, it can be done via bitcoin anonymously insh'allah. Here is the wallet address those that want to help out can send to.

1D2tUhZJVMgiB7dVRgYtA1s7ZsLAuwqHL7

Provided by **MEMRI**

Post on Telegram: "If you know anyone that wants to help support families here fisabillah [for Allah's sake], it can be done via bitcoin anonymously insh'allah."

Further, on January 22, 2019, the Sadaqa Al-Khair channel on Telegram, which says it provides aid to those in need in Syria, shared its Bitcoin address, to which donations could be sent. The channel, which posts content in English and German, highlights the plight of refugees in Syria while repeatedly calling for donations. The Bitcoin announcement included the group's contact number on Telegram and WhatsApp, which has a Turkish country code, along with the message: "Your donations might change somebody's world. Donate now..." As of this writing, the bitcoin address shows zero transactions.[8]

Channel Info — Provided by **MEMRI JTTM**

Sadaqa al-Khair
128 members

t.me/Sadaqa19
Link

We help the needy in Syria.
نحن نساعد المحتاجين في سوريا.
Wir helfen den Bedürftigen in Syrien.

Telegram: +90 553 459 71 53
Whatsapp: +90 553 459 71 53
Telegram Admin: @Sadaqa18

Sadaqa al-Khair
Your donation might change somebody's world. Donate now

Telegram: +90 553 459 71 53
Telegram Admin: @Sadaqa18
Whatsapp: +90 553 459 71 53
Bitcoin: 3KhAHDfTuVnHUfRgQfVP4LcV8SHXpN51u7

Contribute to the Project and
Join and Share this Channel.
----
https://t.me/joinchat/

Telegram
Sadaqa al-Khair

Provided by **MEMRI JTTM**

Sadaqa Al-Khair Telegram account.;Sadaqa Al-Khair publishing its bitcoin address (January 22, 2019).

In a February 1, 2019 post on its Telegram channel, a pro-ISIS Gaza-based jihad group called on Muslims to donate using its Bitcoin address to support jihad and to help equip the

The Coming Storm: Terrorists Using Cryptocurrency

mujahideen. It adds: "The campaign's address on the blockchain website is a41d56d7-ec3a4747- where a0c2-07cfd7046a32." However, this is not formatted as a Bitcoin address. This exact announcement was first posted in 2018.[9]

مؤسسة الرابة للانتاج الاعلامى

#بشرى_سارة

إلى كل من تعذر عليه الغزو أو تجهيز غازياً
إلى كل من يقعون تحت قمع الطواغيت وأعوانهم
إلى كل من توقف عن مساعدة ومساندة المجاهدين خوفاً من بطش الكفر وأهلة

يسر إخوانكم في حملة #جهز_غازيا أن تستقبل تبرعاتكم عبر الخدمة الأكبر أمناً , والأكثر سهولة في عصرنا
خدمة العملة الرقمية : (Bitcoin)
فهي تتناسب مع طبيعة وظروف عصرنا الحالي في معظم دول العالم , خاصة في ظل هذه الحرب الشعواء على الأمة الإسلامية عامة وعلى أبنائها المجاهدين خاصة , حيث توفر السرية في تحويل الأموال من وإلى أي دولة حول العالم بطريقة تقنية متوفرة لدى الجميع .

عنوان الحملة على موقع (Blockchain) العنوان : -a41d56d7-ec3a-4747-
a0c2-07cfd7046a32

Provided by
MEMRI JTTM

@alraia_ps   👁 120   1:14 PM

On February 13, a jihadi Facebook user posted his Cash App username, $ghazicash, and asked what others' Cash App tags were. Cash App is a mobile payment service developed by Square Inc for transferring funds via mobile phone; it supports bitcoin transactions. Users can withdraw funds using the Cash App debit card at ATMs, or can transfer it to any local bank account. As of this writing, the service has over 7 million active users. Cash App usernames are in a "$cashtag" format. It is available at the App Store and on Google Play.[10]

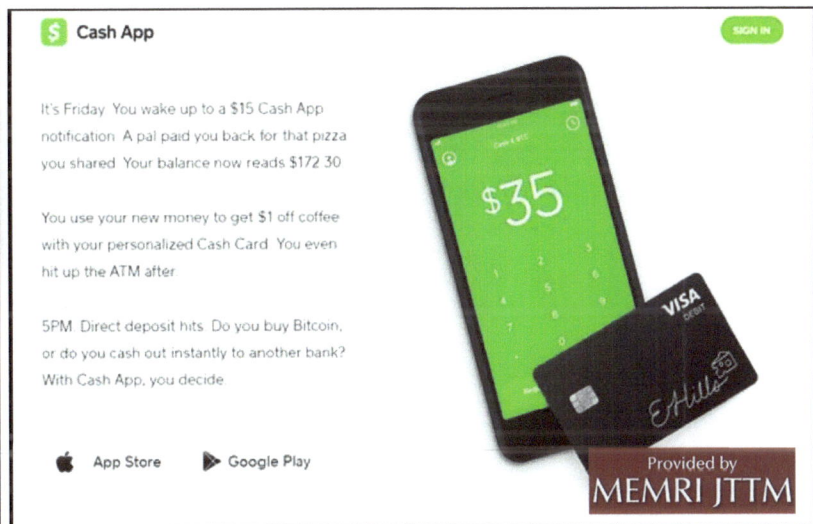

Another example of pro-jihad elements discussing bitcoin and its importance was a video posted May 7, 2019 on the "Daily Reminders" channel on Telegram, New York City-based cleric Muhammad Ibn Muneer. Ibn Muneer maintains a popular YouTube channel on which he has advocated violent jihad and implored Muslims to hate non-Muslims. In this video, he discussed the advantages of cryptocurrencies for the Muslims and urged them to begin using it. He said: "Some aspects of Bitcoin... are closer to the original concept of buying and selling and trading in the shari'a... So some people have the view that Bitcoin is closer to the shari'a than other sources of currency... Another very important point... is if the Muslims don't get involved with the changing fluctuation of the world today, they'll be left behind." He went on to urge Muslims not to be "narrowminded" or "confined in their box of comfortability" while the "kuffar [unbelievers] take it to the next level..."[11]

Daily Reminders
2148 members

Pinned message
Brothers and sisters don't forget to share the channel barkAllah feekom T.me/dailydawah

Daily Reminders
06:28

Provided by
MEMRI JTTM

Mufti Muneer: "Cryptocurrency is the closest to shariah transactions and Muslims should invest in bitcoin and not be late"

While there has been a justified growing chorus of concern from Western government officials about terrorist use of Bitcoin, this concern comes late. As with other online platforms that have allowed terrorists to act freely on them, the longer Bitcoin allows this, the more terrorists will flock to it. Addressing this issue is long overdue.

The Coming Storm: Terrorists Using Cryptocurrency

On social media, known online jihadis are discussing the use of cryptocurrencies such as bitcoin. For example, "S.E.B.," a Canadian who is a prolific sharer of jihadi content on Facebook, including videos of sermons by prominent West-based Islamic extremists such as the late Yemeni-American sheikh and Al-Qaeda figure Anwar Al-'Awlaki, Michigan-based extremist cleric Ahmad Musa Jibril, and Australian pro-ISIS cleric Musa Cerantonio, and is Facebook friends with well-known individuals in jihadi circles, wrote on November 23, 2018: "who knows about buying bitcoins?" In the comments section, he wrote, following a comment about bitcoin having lost "an incredible amount of value in the last year," that "its not for investing."[12]

Facebook user "D.A.," who is thought to live in the Washington, D.C. area, frequently posts content supporting prominent jihadi figures including Al-Awlaki, pro-ISIS sheikh Abdullah Al-Faisal, and Michigan-based Ahmad Musa Jibril. He is Facebook friends with notable ISIS supporters, and his Facebook posts include support for ISIS, discussion of police surveillance, and his interest in using Bitcoin. He wrote on Facebook on December 16, 2017:

"Every kaffir I see now can't stop talking about bitcoin. They are obsessed because they think it is an easy route to happiness… meanwhile Muslims use it to better the ummah and help perform *dawah* [preaching]. Just another example of us vs them."[13]

The same day, D.A. wrote: "For brothers on the *haqq* [truth], is there any interest in starting a thread to discuss does and don'ts of bitcoin? I have a very basic understand and I'm sure many have a much great level of expertise. Just thinking it would be good so Muslims understand how to be safe with it because there are many scammers out there who try to steal and use a lack of understanding to harm hard working Muslims."

"J.M.," from Columbus, Ohio, is Facebook friends with many jihadis, and has discussed the use of cryptocurrencies with them. He has also posted antisemitic content. In 2007, he was sentenced to five months in prison and three years' probation for falsely claiming to be a military veteran; he had gained

fame and risen to prominence in the antiwar movement by claiming that he and other U.S. soldiers had killed civilians in Iraq.[14] He wrote on November 19, 2017: "Who knows about bitcoin that can put me on? I got a bitcoin wallet. I want to start small monthly investments in it. How would I go about it and is it safe."

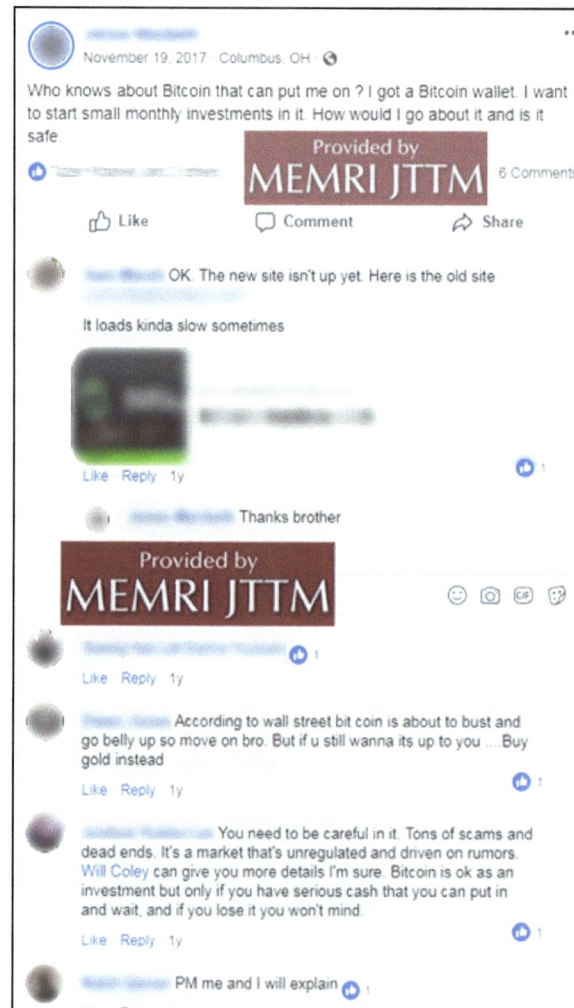

The same day, he wrote: "So I decided not to invest in bitcoin. After some research it is not secure or insured. Its crypto currency so it can be hacked very easily. Apparently stealing other peoples bitcoins is simple and the victims have no recourse for recovery." In the comments section of the post, he said:

The Coming Storm: Terrorists Using Cryptocurrency

"I used to do a bit of hacking myself and know how to backdoor into some one elses system or how to write a script that will look specifically for bitcoin or other crypto currency file markers."

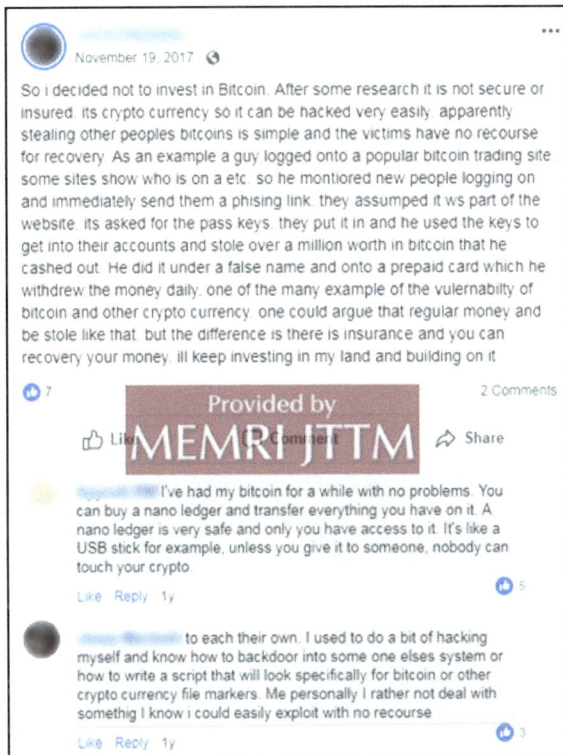

On January 24, 2018, J.M. wrote: "As I said before Bitcoin will crash as governments with fragile economies start to crack down... anything the government can't tax or control they will destroy or make it illegal... I'm glad I researched it it's volitility [sic] and did not invest."

The UK-based "A.M.," who has ties to ISIS and *is* a disciple of extremist UK preacher Anjem Choudary, and whom MEMRI has been monitoring for over three years, wrote on November 24, 2017: "Can someone fill me in about Bitcoins? Are they beneficial in anyway?" One Facebook friend commented on the post: "They're an anonymous crypto-currency, safer for certain things."

A.M. posts prolifically on Facebook, and has over 1,700 Facebook friends, among them Specially Designated Global Terrorist Abu Rumaysah aka Siddhartha Dhar; slain British ISIS militant Abu Abdullah Al-Britani; Australian ISIS fighter Zach Shaam; and Irish convert Khalid Kelley, who reportedly blew himself up at ISIS's directions during the October 2016-July 2017 Battle of Mosul. Facebook friends reported that he was arrested in August 2018, though no news reports of his arrest were found.[15]

On December 16, 2017, "A.Y.," a founder of Invite2Islam (aka InvitcToIslam and Invite to Islam) living in Nottingham, U.K., wrote: "So how do I bitcoin??? Who do I trust ????"

Invite To Islam, an extremist Islamist group based in Birmingham, UK that appears to have ties to ISIS and ISIS fighters and supporters, and on which MEMRI has been reporting since 2014, focuses, inter alia, on radicalizing and converting young people on social media, across many platforms.[16] It shares content such as quotes by *ISIS leader Abu Bakr Al-Baghdadi, Anwar Al-Awlaki and Musa Cerantoni, and other jihadi clerics.*[17] According to MEMRI research in late 2015, the group's Twitter account had well over 6,000 followers, and its Instagram account had over 11,000 followers.

On December 12, 2017, "O.R.," who posts content from Sheikh Anwar Al-'Awlaki, asked: "salaam alaikum: I need to know more about this Bit Coin thing ? LOL down to 1 job and pockets are a bit shabby, is it Haram?" He received several replies.

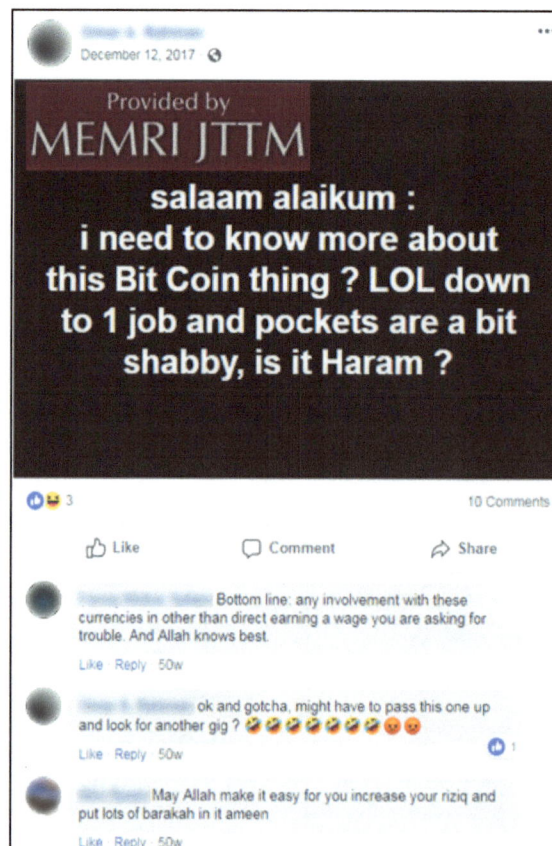

The Coming Storm: Terrorists Using Cryptocurrency

On December 15, 2017, "A.T.," a U.K. man who is Facebook friends with many people who post jihadi content, wrote: "Asalaam Alaikum brothers & Sisters. What are we saying about Bitcoin? Is it Halal?" Commentary attributed to "Shaykh Sulayman Salimullah ar-Ruhayli" followed.

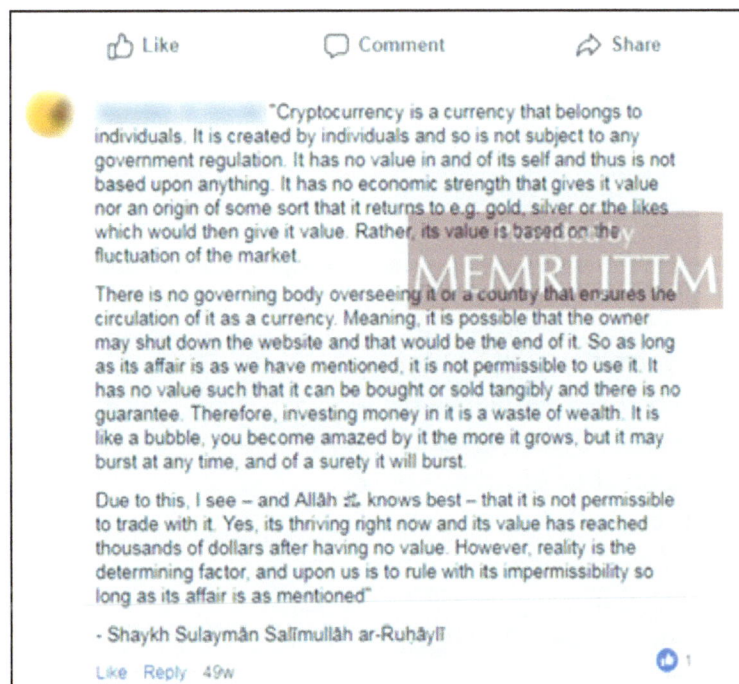

On November 18, 2017, "R.N.," a London woman who is in multiple jihadi cliques active on Facebook, wrote: "Ok, slowly and in plain easy to understand English can someone please explain what bitcoin and crypto currency is and what the whohaa is about it?" A Facebook friend wrote that the transactions are "not logged and no-one can view" them.

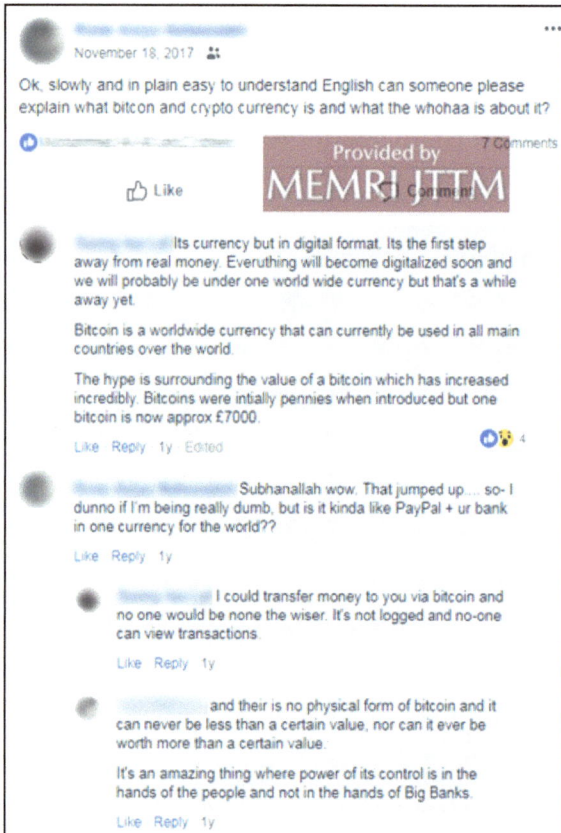

Bosnia-born "A.M.B," currently living in Ontario, Canada, regularly posts on his Facebook page about Islam's incompatibility with democracy, his disgust at Muslim minority groups, and his belief that shari'a law should be implemented everywhere. On December 8, 2017, he posted in the group WE ARE MUSLIMS a meme stating, "What if I told you bitcoin can beat the interest-based banks?" and forwarded a lengthy post from UMMAH Reunited on whether bitcoin was permitted in Islam and whether now was the right time to invest.

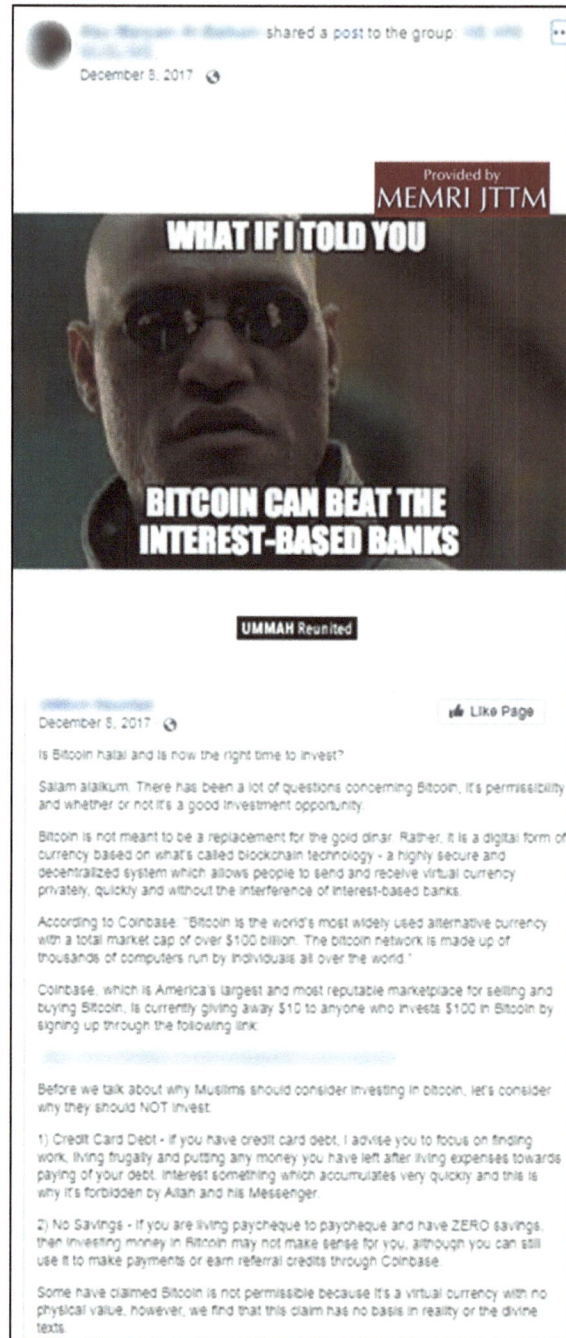

"D.B.," who resides in Coventry, UK, regularly posts anti-LGBTQ content, demands that Muslims not vote, and criticizes Muslims

The Coming Storm: Terrorists Using Cryptocurrency

who are members of the British government on the grounds that they have betrayed their religious community. On February 5, 2014, he posted information about a "Bitcoin Miner."

"D.N.," a Muslim convert living in Oakland Park, Florida, has posted content by antisemitic U.S.-based sheikh Yasir Qadhi, former Finsbury Park Mosque leader Sheikh Abu Hamza, convicted on terrorism charges in the U.K. and U.S. and now serving a life prison sentence in the U.S., and British Salafist convert Abdul Raheem Green. He also reposted on Facebook on December 28, 2018 a tweet by Cryptocom that stated: "Wallet App & MCO Visa Card puts crypto in every wallet. Metal card with no annual fee, unlimited airport lounge access, spending rewards (MCO) and more. Download and reserve today!" and wrote, "I gotta check this out."

Two Years After MEMRI-Assisted Congressional Letter To Telegram CEO Durov Raising Alarm About Upcoming Release Of Platform's Cryptocurrency And Expressing Concern About Extensive Terrorist Use Of His Platform Goes Unanswered, Congressmen Work With MEMRI To Send Follow-Up Letter Demanding Response

On October 25, 2018, Rep. Ted Poe (R-TX) and Rep. Brad Sherman (D-CA), Chairman and Ranking Member, respectively, of the House Subcommittee on Terrorism, Non-Proliferation, and Trade, cosigned a congressional letter (below) initiated by the MEMRI CJL to Telegram founder and CEO Pavel Durov demanding that he take immediate action against the well-known, well-documented widespread terrorist use of his platform. This letter increased pressure on Telegram to address the issue of terrorism on its platform, following an earlier letter to Durov – to which Durov never responded – that also originated with MEMRI in December 2016 and was cosigned by Rep. Poe and Rep. Sherman. Terrorists continue to rely on Telegram to disseminate propaganda, recruit new members, coordinate plans, and fundraise, as platforms such as Facebook and Twitter increasingly crack down on such content. Telegram, exposed by MEMRI in late December 2016 as ISIS's and other terrorist groups' "app of choice" – in a report covered by *The Washington Post* – remains obstinately opposed to taking meaningful action against the extremely extensive terrorist content, particularly ISIS content, on its platform.

Telegram is already a dangerous vehicle for terrorist fundraising; the threat will grow exponentially when Telegram releases its own TON blockchain and Gram cryptocurrency. In 2018, the company raised $1.7 billion in two private initial coin offering rounds for them; on April 11, 2019, it was reported that TON is already in private testing mode, reportedly set for release in the third quarter of 2019. Gram will launch as one of the most valuable cryptocurrencies ever, and every user will automatically receive their own wallet for it. This will be a game-changer for terrorist groups' fundraising and should be the No. 1 focus for counterterrorism officials.

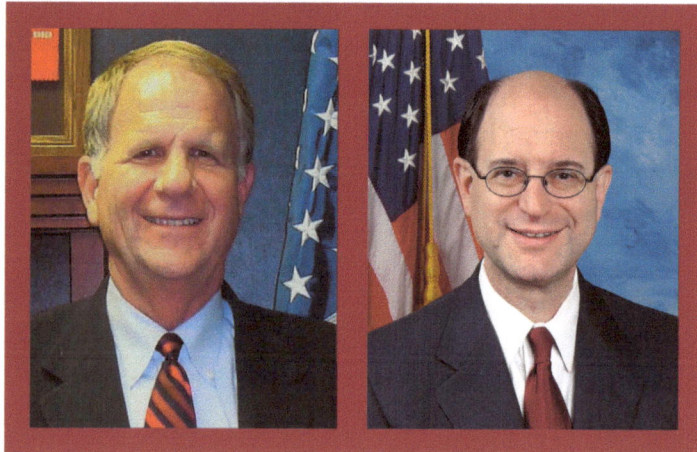

The Coming Storm: Terrorists Using Cryptocurrency

Congress of the United States
House of Representatives
Washington, DC 20515–4302

October 25, 2018

Pavel Durov
Founder and CEO, Telegram Messenger LLP
71-75 Shelton Street
Covent Garden
London
United Kingdom

Dear Mr. Durov and Counsel:

In December 2016, we sent you a letter expressing grave concerns about reports of Foreign Terrorist Organizations (FTOs) and their supporters increasingly utilizing Telegram's platform to incite violence and organize attacks. Unfortunately, that letter was never answered. As Chairman and Chairman Emeritus of the House Foreign Affairs Committee's Subcommittee on Terrorism, Nonproliferation, and Trade, we remain deeply concerned about these illicit uses of your platform, but also by your failure to acknowledge and address them in response to an official congressional inquiry.

Almost two years later, the problems discussed in our letter appear to have significantly worsened. Just in the last few months alone, we have seen jihadists use Telegram to claim responsibility for attacks in Paris, France, and Liege, Belgium and praise their perpetrators.

It is well documented that violent extremists are finding a home on Telegram and remaining there. Although encrypted services can be valuable tools for privacy and countering censorship, we must also recognize that others use Telegram to disseminate propaganda, drive terrorist fundraising, recruit new members, and coordinate attacks. Some channels encourage "lone-wolf" attacks by sharing tutorials on terror tactics, such as bomb-making, conducting stabbing attacks, target selection, and methods to avoid detection by security services. A recent ISIS-linked campaign on your service called on Muslims to move to Afghanistan to wage jihad. ISIS affiliates have even used Telegram to distribute "kill lists" with the names, addresses, and other personal details of U.S. government personnel, police officers, and employees of major U.S. companies. Other terrorist organizations such Al-Qaeda in the Arabian Peninsula, one of the world's most lethal terrorist groups, as well as Hamas, Hezbollah, and the Taliban have also adopted Telegram to disseminate propaganda. Despite these dangerous trends and increased vigilance by other social media platforms, channels such as the ISIS-affiliated Al-Nashir news channel continue to encourage attacks on U.S interests over Telegram with impunity.

With Telegram's launch of its Telegram Open Network (TON) blockchain and Gram cryptocurrency, there is additional concern that such applications will make it even easier for terrorists to fundraise without disruption. Terrorist fundraising is already a persistent problem on

Telegram due to the anonymity and privacy that the platforms provides. Groups such as the ISIS-affiliated Ibn Taymiyyah Media Center (ITMC) host elaborate fundraising campaigns on your platform. Through an online campaign called "Jahezona," ("Equip Us"), launched in 2015, ITMC raised funds for Salafi jihadi organizations in the Gaza Strip by urging supporters to donate money to militants as a required religious duty. The Syrian jihadist group Ha'yat Tahrir Al-Sham (HTS) has a campaign underway on Telegram to raise funds, stating that preparing a suicide bomber costs $20,000 and carrying out a deep-strike operation costs $1,500. Earlier this year, the al-Qaeda-linked English-language magazine Al-Haqiqa, which is distributed over Telegram, published an article stating "we see lots of potential for the use of cryptocurrencies for our purposes." We believe that given Telegram's popularity among known terrorist groups, its use as a terror financing tool will only increase unless immediate steps are taken.

As our previous letter noted, Telegram need not face this challenge blind or alone. We hope Telegram learns from other platforms that have reduced extremism on their platforms while still maintaining an emphasis on privacy. For example, other platforms have made it easier for users to report terrorist content, introduced algorithms and other automated mechanisms to assist manual reviews and remove accounts promoting violence, trained review teams on what to look for and ways terrorists abuse their platforms, and, more recently, collaborated on a shared database of terrorist content. We also suggest that you work with the Global Internet Forum to Counter Terrorism, which is a new tech industry consortium pooling resources and know-how to confront terrorists online.

As Members of Congress, we are strong advocates for the right to privacy and appreciate Telegram's strong commitment to protecting this right. We also appreciate the power of encrypted technologies to undermine censorship and enable free speech in certain critical contexts. However, when a company's services are being used to promote and facilitate terrorist activity that results in the deaths of innocent individuals, we have an obligation to speak out. We acknowledge that your "ISIS Watch" application does claim to have removed thousands of terrorism-linked accounts, however we do not believe—given the examples listed above—that this has sufficiently stopped extremist activity on your platform. We urge you to take additional steps to curtail jihadist content on Telegram.

We ask that you respond to this letter with immediate urgency specifying a plan of action to combat extremist content on Telegram and create safeguards to prevent terrorist groups from using the platform as a secure fundraising tool. This includes a detailed summary of what measures you will undertake to ensure the addition of cryptocurrency functionality on your service will not facilitate money laundering or terrorist fundraising efforts. Without credible and verifiable steps taken within your company, we may be compelled to pursue other measures to ensure the security of Americans and our allies. Recently, a number of technology platforms have been called to testify under oath before Congress, so we hope you appreciate the keen interest of the American people in preventing the illicit use of online services. Again, we urge you and your company to take immediate action.

We look forward to your cooperation on this important matter.

Sincerely,

Ted Poe
Chairman, Subcommittee on Terrorism, Nonproliferation, and Trade

Brad Sherman
Ranking Member, Subcommittee on Asia and the Pacific

The Coming Storm: Terrorists Using Cryptocurrency

The Imminent Release Of Telegram's Cryptocurrency, ISIS's Encryption App Of Choice – An International Security Catastrophe In The Making

On April 2, 2018, the MEMRI Cyber Jihad Lab published the following report, titled The Imminent Release Of Telegram's Cryptocurrency, ISIS's Encryption App Of Choice – An International Security Catastrophe In The Making.

ميمري
I&A #1387
The Imminent Release Of Telegram's Cryptocurrency, ISIS's Encryption App Of Choice – An International Security Catastrophe In The Making

Steven Stalinsky | March 30, 2018

Preface

Nearly 17 years after the 9/11 attacks and the subsequent 9/11 Commission Report recommendations, a Telegram cryptocurrency will shatter all the lessons learned about terrorism financing, and will be used to fund terrorism worldwide. The recommendations stated: "Vigorous efforts to track terrorist financing must remain front and center in U.S. counterterrorism efforts. The government has recognized that information about terrorist money helps us to understand their networks, search them out, and disrupt their operations. Intelligence and law enforcement have targeted the relatively small number of financial facilitators – individuals Al-Qaeda relied on for their ability to raise and deliver money – at the core of Al-Qaeda's revenue stream."

As terrorist groups such as the Islamic State (ISIS) have grown more sophisticated and do much of their fundraising online, it has become almost impossible to follow these recommendations. Telegram's introduction of its own cryptocurrency will allow terrorist groups to easily distribute money, raise funds, and so on. The question is: Can anything be done to stop it? The 9/11 commission report detailed how fundraising for the 9/11 attacks relied on imams and mosques, through hawala (an informal trust-based system of transferring funds) and other fairly primitive means. Cryptocurrencies have already made terrorist fundraising more difficult to track, and with Telegram's and Telegram CEO Pavel Durov's history, the problem will increase by orders of magnitude.

Introduction

A white paper "leaked" in early January detailed how the encryption app Telegram aims to join the cryptocurrency craze by adding a "third generation" blockchain with more advanced capabilities than Bitcoin and other existing cryptocurrencies with a mainstream appeal. Pavel Durov, who seeks to raise over a billion dollars for the development of the Telegram Open Network (TON) Blockchain, whose representative cryptocurrency is Gram, and for "the development and maintenance of Telegram Messenger, and other purposes," has already gained the interest of wealthy Internet investors, and has won over many in the leading economic media to his plans.

Following an initial coin offering (ICO), Telegram is on track to raise that billion dollars, reportedly in just four months. As of March 2018, Telegram had taken in $850 million from wealthy venture capital firms, and it reportedly aims to raise another $850 million over the next month and a third $850 million in the round after that, according to documents associated with the offering.[18] Media outlets, including Bloomberg, state that $2.5 billion could be raised. It should be noted that Durov, who, according to *Forbes,* already has a net worth of $1.7 billion and who created the popular Russian social media platform VK, which is widely used by Russian-speaking jihadis, could see his personal wealth increase dramatically.

Most media reports about the Telegram ICO have either failed to disclose or have downplayed the fact that the Telegram app is the heart and soul of cyber jihad, and is hugely popular with terrorists, particularly with ISIS. Since 2015, when it was used in the November 2015 Paris attacks, it has emerged as jihadis' preferred app for encrypted communications, including planning and carrying out attacks, as arrests worldwide have shown. On March 15, 2018, authorities arrested a Cuban man in Bogotá who had planned to bomb a restaurant frequented by U.S. diplomats and who had on Telegram communicated his desire to blow himself up for the Islamic State. Just last week, on March 23, ISIS used Telegram to claim responsibility for its attack in the southern French town of Trèbes (see image below).

Previously based in Germany and London, and registered as a British company, Telegram has tried "a number of locations." Noting that the company is currently in Dubai, Pavel Durov said on Twitter that "we don't consider it our permanent base," but the company is there until "local regulations change." Durov boasts about moving Telegram from place to place to evade government oversight and restrictions over the past two years, and his tweets have been datelined N.Y., U.K., France, Switzerland, and other locations throughout the world.

The Coming Storm: Terrorists Using Cryptocurrency

It should, however, be noted that Telegram has servers throughout the West, including in the U.S.

Pavel Durov @durov
Replying to @peterdobey @telegram
We are currently based in Dubai, but we don't consider it our permanent base. We are unlikely to ever consider any location to be our permanent base.
6:09 PM - 16 Jan 2018 from London, England
25 Retweets 52 Likes
11   25   52

Durov often acts as if he is above the law and deserves immunity from criticism about the terrorists on his platform. Governments worldwide – first the U.S., then the U.K., France, Germany, and others in Europe – have also warned about Telegram as a hub for terrorist planning, recruitment, and daily operations and communications.

Should Telegram launch its own cryptocurrency, it will surely become the go-to financial apparatus for terrorist groups – and at the same time create an impossible situation for counterterrorism officials trying to stop terrorist financing. Were no lessons learned from the 9/11 Commission Report section on terrorist fundraising?

In a possible sign of trouble ahead, mounting concern about bad elements investing in Telegram's cryptocurrency have led the company to take some initial precautions. In mid-March, it was announced that individuals on the sanctions lists of the U.S., U.K., European Union, or United Nations, including Iran, will not be allowed to take part in Telegram Open Network initial coin offerings (ICOs), and will be forbidden by the official Telegram group investor agreement from buying Gram cryptocurrency, according to the Russian news agency RBC. The ban extends to all residents of the Crimean Peninsula, Cuba, Iran, the Democratic People's Republic of Korea, and Syria.

In December 2016, MEMRI released its seminal study Germany-Based Encrypted Messaging App Telegram Emerges As Jihadis' Preferred Communications Platform. A year later, Telegram is even more widely used by terrorist organizations. Adding its own cryptocurrency will further increase both its popularity and its use by terrorists – particularly since Telegram has not even begun to seriously address terrorist use of its platform.

On December 26, 2016, one day after MEMRI exposed Telegram as ISIS's and other jihadis' "app of choice" – as described by *The Washington Post* on its front page on December 24 – Telegram responded by creating its ISIS Watch channel/bot. In daily updates, the account bot notes the number of ISIS accounts Telegram claims to have removed in the previous 24 hours – a number totaling 5,000-8,500 monthly. But Telegram offers no proof that this actually happens, and as an observer of Telegram and of terrorist groups online, I can attest that many accounts that do disappear soon return.

Provided by MEMRI JTTM

Pavel Durov @durov

Speaking of ISIS: over 660 public ISIS channels banned since November; 5-10 removed daily following reports to abuse@telegram.org

1:01 AM - 15 Jan 2016

203 Retweets 262 Likes

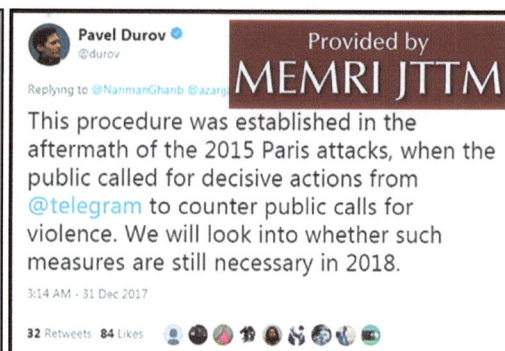

Provided by MEMRI JTTM

Pavel Durov @durov

Replying to @NanmanGhanb @azar...

This procedure was established in the aftermath of the 2015 Paris attacks, when the public called for decisive actions from @telegram to counter public calls for violence. We will look into whether such measures are still necessary in 2018.

3:14 AM - 31 Dec 2017

32 Retweets 84 Likes

On January 31, 2017, following reports that Telegram was used in the November 2015 Paris attacks, Durov had the audacity to tweet on the matter of Telegram's procedure for removing terrorist content: "We will look into whether such measures are still necessary in 2018." A recent post on Telegram of "guidelines for doing just terror operations" shows how to lethally stab someone and what kind of knife to use, advising throat slashing.

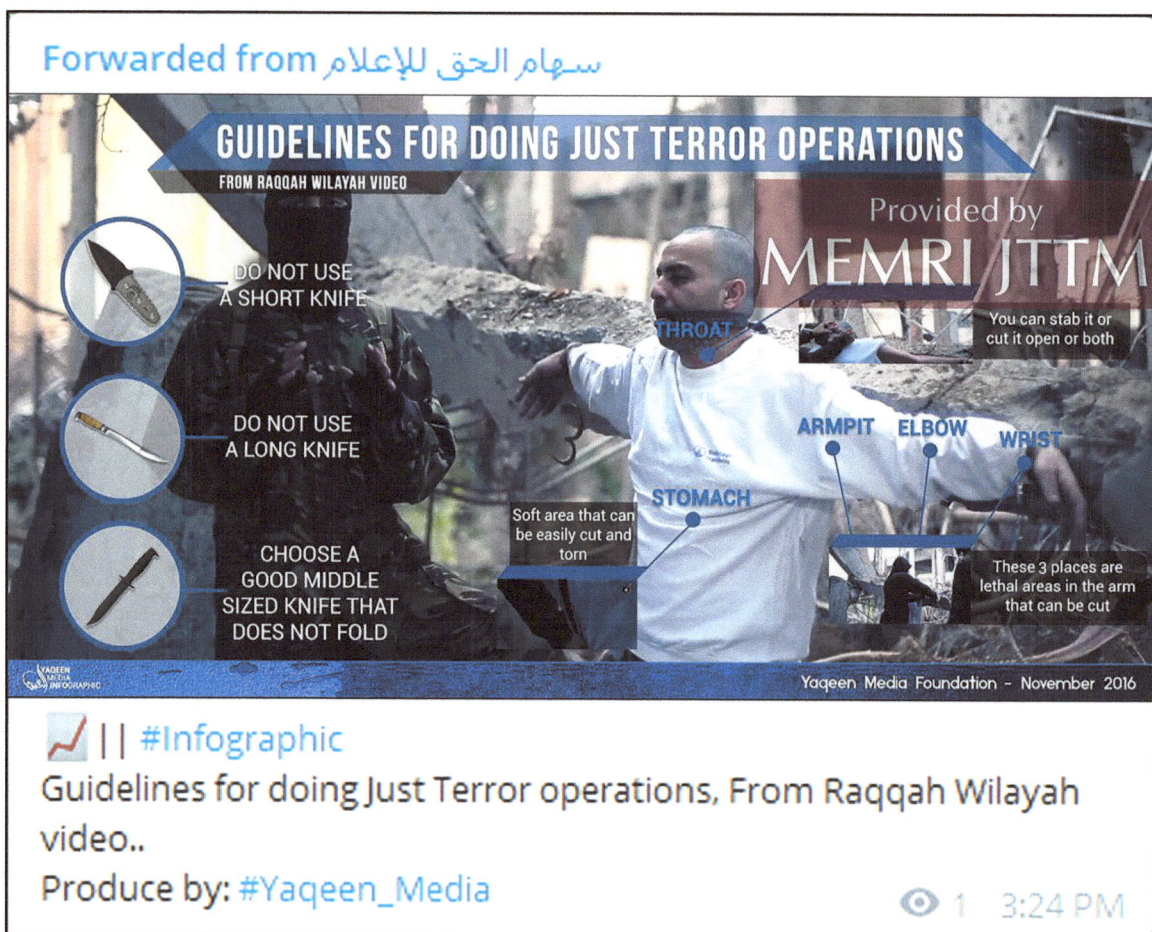

All of the methods of killing described in this official ISIS video, a screenshot from which is shared on Telegram, are carried out on a living man; he dies on camera.

The Coming Storm: Terrorists Using Cryptocurrency

Another post this week on the LM WORLDWIDE (Lone Mujahideen) Telegram Channel included "tips for the mujahideen in the enemies' lands," gives specific instructions on what to do before planning, during planning, and in the implementation of an attack, including "strike violently and try to inflict the greatest losses in the enemy" and "keep killing until you are killed."

The following report explains why a Telegram cryptocurrency is a national security threat, detailing the justified concerns of the U.S. and other Western governments about current terrorist use of Telegram and how and why governments have not been able to successfully pressure the company to tackle it. It will also explain how this is going to change with the coming ICO and how terrorists already freely use Telegram for fundraising. The report will offer recommendations for how the U.S. government and Western companies can finally get Telegram to remove terrorist activity from its platform and what measures should be taken if Telegram does not comply.

Governments Are Concerned About Terrorist Use Of Cryptocurrencies – But May Be Unaware of The Looming International Security Danger Of Telegram Cryptocurrency

As the value of Bitcoin reached new highs earlier in the year and garnered increased media and other attention, there has been heightened concern from governments worldwide that terrorist groups are using digital wallets to send and receive funds using Bitcoin and other cryptocurrencies. At the same time, as ISIS's physical caliphate crumbles, along with its own once-heralded economy that included its own currency and banks, it is naturally turning to cryptocurrencies. In recent months, French, Italian, Australia, Japan, U.K., and other legislatures worldwide have sought more regulations on Bitcoin and other cryptocurrencies because of potential terrorist usage. In the U.S., there have been bipartisan Congressional hearings and legislation on the matter including in January 2018, when Rep. Ted Budd (R-N.C.) introduced H.R.4752, the Financial Technology Innovation and Defense Act, to establish a task force to combat terrorist use of cryptocurrency. These concerned government bodies worldwide may be unaware of the looming international security danger associated with a Telegram cryptocurrency.

Western Governments Have Not Had The Leverage To Pressure Durov – Until Now

Inexplicably, although Telegram is a major highway for the daily activity of ISIS and other terrorist groups, and has been at the center of multiple terrorist cases worldwide, there has been no significant pressure on Telegram to remove terrorist content. Reasons for the lack of pressure could be that the location of Telegram's headquarters is not available, and Durov's international movements, and those of the company's assets, make it difficult to compel the company to tackle its terrorism problem. The new official cryptocurrency changes all that. As other social media companies, led by Facebook, make tremendous efforts to fight terrorist content online, Durov remains defiant about his platform's popularity among terrorists worldwide, and his responses to accusations are filled with half-truths and snark.

The Coming Storm: Terrorists Using Cryptocurrency

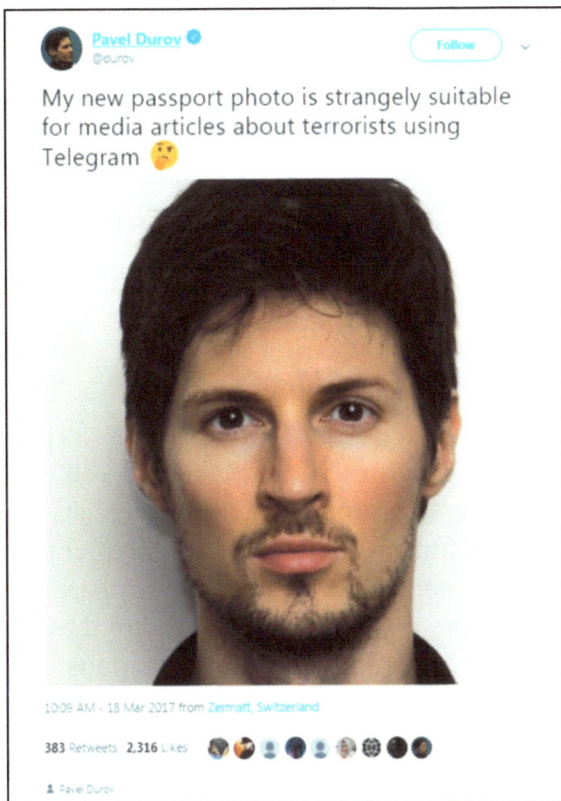

Pavel Durov @durov · Follow

My new passport photo is strangely suitable for media articles about terrorists using Telegram 🤔

10:09 AM - 18 Mar 2017 from Zermatt, Switzerland

383 Retweets 2,316 Likes

Pavel Durov

Despite Durov's unsubstantiated claims that Telegram has been removing terrorist content, he has been consistently arrogant in the face of Western government pressure to do so. The day of the March 18, 2017 Orly Airport attack – the same week as the Westminster attack in London – he tweeted that his "new passport photo is strangely suitable for media articles about terrorists using Telegram."

His attitude was best summed up by the *Evening Standard*'s James Ashton: "As governments wrestle with social media's role in tracking criminals and censoring content, Durov has become a divisive figure akin to Julian Assange or Edward Snowden. Telegram was one platform on which Islamic State plotted its [November 2015] attack on the Bataclan [theater in Paris]."

Pavel Durov @durov · Follow

Some media claimed Telegram did little or nothing to stop ISIS from using its platform. Nothing can be further from the truth!

Telegram Messenger @telegram
Every day we block over 60 ISIS-related channels before they get any traction, more than 2,000 channels each month. telegram.me/ISISwatch

9:04 AM - 27 Dec 2016

166 Retweets 300 Likes

Pavel Durov @durov · Follow

Our policy is simple: privacy is paramount. Public channels, however, have nothing to do with privacy. ISIS public channels will be blocked.

10:41 AM - 19 Nov 2015

174 Retweets 335 Likes

Governments of Western countries, including Germany, France, the U.K., the U.S., and others, have all sounded alarms about terrorists on Telegram. For example, on August 23, 2016, it was reported that France and Germany were pushing for Europe-wide rules requiring the makers of encrypted messaging apps, notably Telegram, to help governments monitor communications among suspected extremists. At a news conference with German Interior Minister Thomas de Maiziere, French Interior Minister Bernard Cazeneuve said that French investigators, armed with a court order, had been unable to even contact "an interlocutor" at Telegram. Durov wrote, in response to a question submitted via his platform about Cazeneuve's statement: "We haven't received any such request and have no idea what the French officials are after. In any case, Telegram Secret Chats and information on them are not logged on our servers." Additionally, Mounir Mahjoubi, president of the National Digital Council, an independent advisory group in France established by French president Nicolas Sarkozy that focuses on privacy issues, said: "There is an issue with Telegram. They have done everything to make it a technological nightmare to find where their server is."

More recently, in her speech at the Davos Forum in January 2018, U.K. Prime Minister Teresa May stated: "Technology companies still need to do more in stepping up to their responsibilities for dealing with harmful and illegal online activity." While big tech companies need to do this, she added, it is also important for small companies to do so as well: "Smaller platforms can quickly become home to criminals and terrorists... We have seen that happen with Telegram. And we need to see more co-operation from smaller platforms like this... We also need cross-industry responses... No one wants to be known as 'the terrorists' platform' or the first-choice app for pedophiles."

Terrorist Fundraising On Telegram

ISIS, Al-Qaeda, and other jihadi groups have been using and experimenting with cryptocurrencies for a few years now, for fundraising and planning attacks, sharing and authoring articles about them, and preparing for when enough terrorist groups and followers become comfortable with them and use them widely to transfer funds and make them a regular part of their terrorist activities – as regular as using social media is today, especially platforms such as Telegram. Terrorist fundraising is already underway on Telegram – as are the dissemination of tips on how to do so, direct requests for funds for purchasing weapons, and more.

Last month, the pro-Al-Qaeda English-language magazine Al-Haqiqa, which is distributed on Telegram, included an article examining the shari'a permissibility of using Bitcoin and similar currencies to fund jihad. It stated: "We see lots of potential for the use of cryptocurrencies for our purposes."

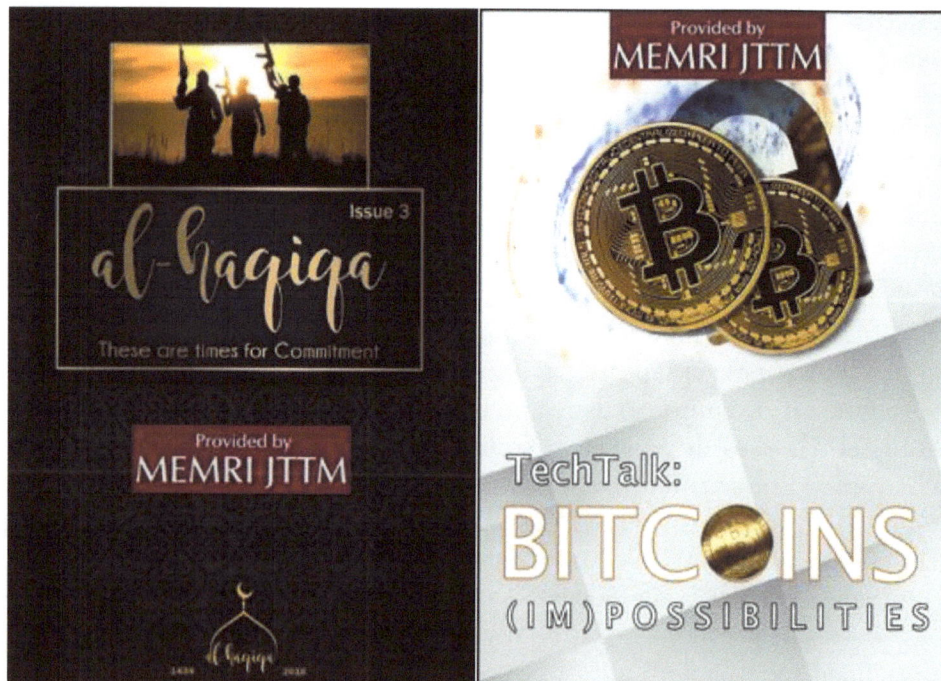

Cover of Al-Haqiqa with article on cryptocurrency.

The Coming Storm: Terrorists Using Cryptocurrency

Forwarded from بِسْمِ اللَّهِ الرَّحْمَنِ الرَّحِيم Islamic Quotes **TM** (Back Up)
Al-Haqiqa Magazine brings you news and background articles about Sham and other parts of the world.

This group is created mainly to spread upcoming issues of our magazine, Insh'Allah.

You can contact our Admin : @AlHaqiqa

Group link : @alhaqiqamagazine

Provided by
MEMRI JTTM

Al-Haqiqa distribution on Telegram.

The pro-ISIS Ibn Taymiyyah Media Center (ITMC), identified with the Salafi-jihadi stream in the Gaza Strip, has been conducting an online campaign called "Jahezona" ("Equip Us") to raise funds for the Salafi jihad organizations in Gaza since 2015. The campaign, whose motto "The money will come from you and the blood will come from us" calls upon Muslims to donate funds for the Gazan mujahideen, and also to assist by spreading the word about the campaign. It stresses that the mujahideen are in dire need of such aid and that "waging jihad by means of money" (i.e. financially assisting the jihad fighters and their families) is an important religious duty incumbent upon each and every Muslim. The campaign also emphasizes that those donating money for jihad are like those fighting with their own hands, and that the religious texts and scholars "even put this duty before" the duty of waging jihad on the battlefield. Other, similar campaigns are also underway, notably one relaunched nearly every year by the Gaza-based Salafi-jihadi group Jaysh Al-Ummah.

The following are items from the campaign posted this month on the group's Telegram channel devoted to fundraising. The graphic below says that the donated money will be spent on: waging jihad for the sake of Allah; training the mujahideen and equipping them with weapons and ammunition; manufacturing weapons such as rockets, bombs, and IEDs; training the mujahideen and developing their skills; teaching and preaching Islamic education; jihadi media; raising the level of security alertness for the mujahideen and for securing and protecting those on the run; and social services assisting the families of the martyred and imprisoned mujahideen.

The most important organization to openly use Telegram for fundraising is the Syria-based anti-Assad Al-Sadaqah organization, which promotes itself as itself as "an independent charity organization that is benefiting and providing the Mujahidin in Syria with weapons, financial aid, and other projects relating to Jihad." Its ongoing fundraising campaign on social media, especially on Telegram, is aimed at Americans and other Westerners — and a growing number of them are donating, disseminating its posts, and chatting about how to donate. The campaign shares its Bitcoin wallet number on its Telegram channel, which it opened on November 8, 2017. Its image and video on it has text on the bottom of its pages "Donate anonymously with cryptocurrency."

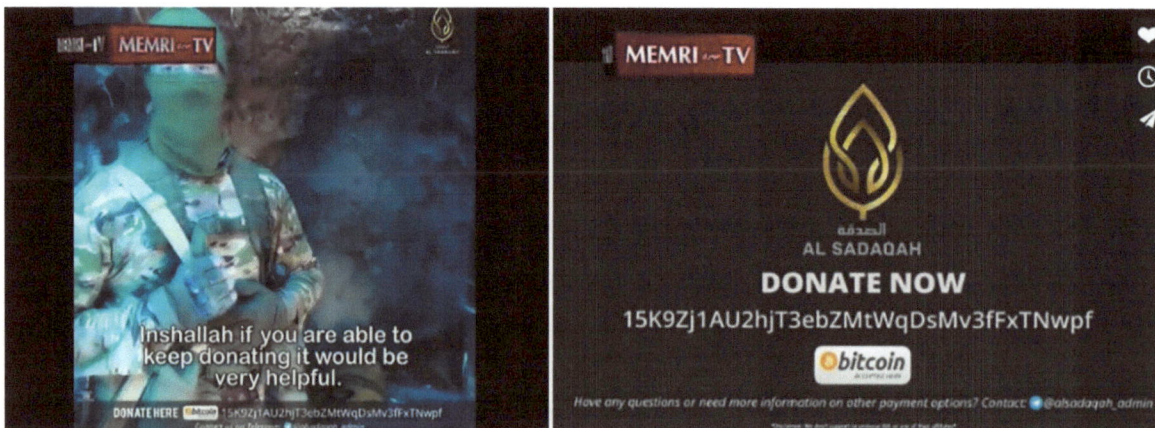

Images from the Al-Sadaqah Telegram channel.

The Coming Storm: Terrorists Using Cryptocurrency

On the pro-Hay'at Tahrir Al-Sham (HTS) Telegram channel "HTS Correspondent" a campaign is underway calling on Muslims to donate money to support the mujahideen waging jihad against the polytheists.[19] The channel also states that preparing a suicide bomber for a mission with a car bomb costs $20,000, while preparing one commando to carry out a deep-strike operation costs $1,500. The list on the right below details more military expenses: preparing a fighter for a storming operation costs $1,000; a mortar shell costs $15; a mortar tube costs $600; an Al-Feel rocket costs $900; a mountain motorbike costs $900; and a pickup truck costs $10,000.[20]

Another of the many examples of fundraising going on right now on Telegram is a call by a well-known pro-ISIS disseminator for Muslims to donate money for ISIS and the caliphate. It states in part: "Donate your wealth to spread and defend the creed and methodology of the Islamic State, to defend the Mujahideen in the battlegrounds, and to defeat unbelief, polytheism, and the filthy rejectionists [Shi'ites]... O monotheists, hasten and make *hijra* [emigration] to the battleground of jihad and the fight to defend Allah's religion and the mujahideen [who are defending] Allah's religion in the Islamic caliphate. If you defend Allah, Allah will grant you victory until you become among those who fight for His sake steadfastly. If you are unable to be in the battlegrounds of jihad for legitimate reasons, be aware that you are still obligated to wage jihad which has levels. One of it is to wage jihadi with your wealth. O you who are not in the battleground of jihad, can't you spend and give your wealth in the sake of Allah?"

The Coming Storm: Terrorists Using Cryptocurrency

How The Planned Telegram Cryptocurrency Will Work

A promotional video for the Telegram Open Network (TON), posted on YouTube on December 22, 2017, promises that the "fast, scalable, and user-friendly" cryptocurrency and blockchain platform will "revolutionize blockchain technology." All Telegram users, it says – 180 million, with 500,000 more joining every day, and 70 billion messages delivered daily – will get a TON wallet, making it "the world's most adopted cryptocurrency."

"A blockchain Telegram Open Network (TON) and a crypto-currency called Gram will upgrade Telegram from just a messaging service, making it also a payment system that will compete with Visa and MasterCard as a mainstream payment network. The new network aims to include a digital wallet, storage services similar to those of Dropbox Inc., a proxy layer to hide identity, an applications platform, and a payments channel."

"With its ability to handle millions of secure transactions per second" due to its "unique multi-blockchain architecture," all blockchains can quickly exchange data with a "smart routing system." TON, it says, uses "direct payment channels to transfer banking in milliseconds" and can "easily accommodate billions of users." The video concludes: "In 2018, 200 million Telegram users will get access to TON, and its fast-user-friendly cryptocurrency and app platform."

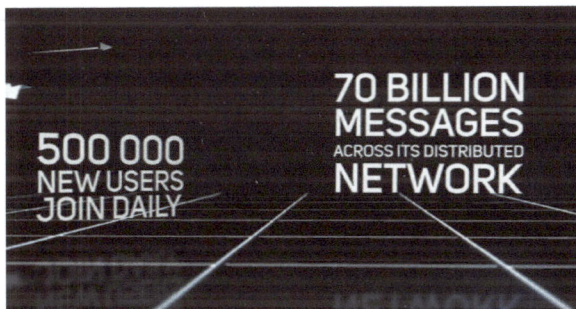

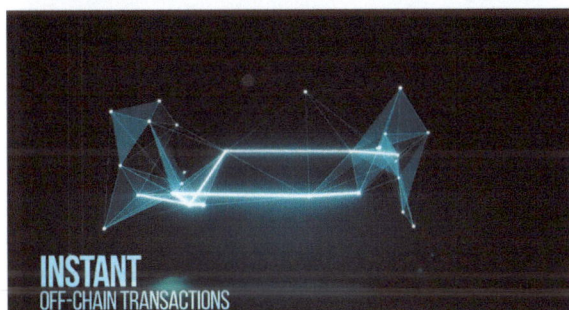

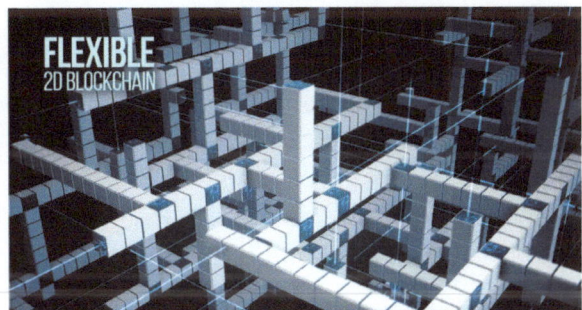

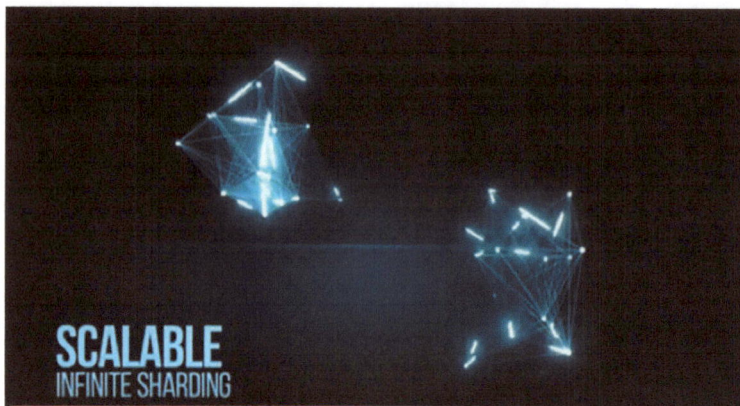
SCALABLE
INFINITE SHARDING

Over The Past Two Years, MEMRI Has Supported And Provided Research For Efforts To Pressure Durov On Capitol Hill – Now With Telegram's Cryptocurrency, Further Urgent Action Is Needed

Since the summer of 2016, MEMRI has been meeting with government agencies and offices on Capitol Hill on the issue of Telegram being used by terrorist groups worldwide. We observed the platform as a handful of terrorist accounts quickly exploded into thousands, and we sounded the alarm. Shortly thereafter, Telegram was found to be connected to major terrorist attacks.

On December 16, 2016, Rep. Ted Poe, Chairman of the House Subcommittee on Terrorism, Nonproliferation and Trade, and Rep. Brad Sherman, the Committee's Ranking Member, called Durov out in a Congressional letter that included exclusive MEMRI research. MEMRI has worked for years with these two members of Congress on the issue of removing terrorist content from social media platforms, including, most recently and specifically, Telegram.

The letter reads: "We write to communi-cate our grave concerns about reports regarding the recent migration of Foreign Terrorist Organizations (FTOs) and their supporters from other social media platforms to Telegram. Terrorists are reportedly using Telegram's encrypted service to disseminate propaganda, drive fundraising, recruit new members, and coordinate attacks... It is our hope that Telegram will learn from other social media companies who have successfully maintained their emphasis on privacy while reducing the terrorist footprint on their platforms. They did this by making significant investments over a sustained period of time. These investments included making it easier for users to report terrorist content, having dedicated teams to examine reports and proactively search for terrorist content, using algorithms and other automated mechanisms to assist manual reviews, quickly identifying and removing accounts promoting terrorism, regularly training review teams on what to look for and new ways terrorists are using the platform, and, most recently, coming together to form a shared database of terrorist content." It continued: "[N]o private company should allow its services to be used to promote terrorism and plan out attacks that spill innocent blood."

After noting that "[h]undreds of channels affiliated with ISIS and other terrorist organizations still find refuge in Telegram's encrypted service," it enumerated several examples of this content, and concluded: "We respectfully request that you do all in your power to prevent terrorists from exploiting Telegram to advance [their] lethal cause."

The Coming Storm: Terrorists Using Cryptocurrency

Congress of the United States
Washington, DC 20515

December 16, 2016

Pavel Durov
Founder and CEO
Telegram

Dear Mr. Durov:

We write to communicate our grave concerns about reports regarding the recent migration of Foreign Terrorist Organizations (FTOs) and their supporters from other social media platforms to Telegram. Terrorists are reportedly using Telegram's encrypted service to disseminate propaganda, drive fundraising, recruit new members, and coordinate attacks.

As U.S. Members of Congress, we are strong advocates for the right to privacy and appreciate Telegram's strong commitment to protecting this right. However, no private company should allow its services to be used to promote terrorism and plan out attacks that spill innocent blood. So we were thankful to see your removal of 78 ISIS-related channels across 12 languages on the heels of last year's attacks in Paris. This, however, is just a start.

Hundreds of channels affiliated with ISIS and other terrorist organizations still find refuge in Telegram's encrypted service. "Nashir News Channel", an outlet of ISIS that delivers pro-ISIS news in several languages and has been suspended on Twitter and YouTube, attracts well over 10,000 regular followers on Telegram. Other ISIS channels have issued calls to target U.S. military bases in the Middle East and facilitated the selling of two prisoners, one from Norway and another from China. "Kill lists" distributed this past summer by ISIS on Telegram included names, addresses, and other personal details of hundreds of U.S government personnel, police officers, and employees of major U.S. companies. Al-Qaeda in the Arabian Peninsula (AQAP), one of the most lethal terrorist groups in the world, uses Telegram to release its statements and propaganda. Other terrorist groups such as Hamas, Jabhat Al-Nusra, and the Taliban are also on Telegram. We are troubled at these examples and many others that illustrate how your platform is allowing the spread of terrorist ideology and the coordination of actual terrorist attacks.

Telegram is not the first to face this challenge. It is our hope that Telegram will learn from other social media companies who have successfully maintained their emphasis on privacy while reducing the terrorist footprint on their platforms. They did this by making significant investments over a sustained period of time. These investments included making it easier for users to report terrorist content, having dedicated teams to examine reports and proactively search for terrorist content, using algorithms and other automated mechanisms to assist manual reviews, quickly identifying and removing accounts promoting terrorism, regularly training review teams on what to look for and new ways terrorists are using the platform, and, most recently, coming together to form a shared database of terrorist content.

Terrorists will continue to try to kill innocent people even if social media companies never existed, but it is in the interest of us all to not enable and even support their nefarious actions. We respectfully request that you do all in your power to prevent terrorists from exploiting Telegram to advance its lethal cause.

Sincerely,

Ted Poe
Chairman, Subcommittee on Terrorism,
Nonproliferation, and Trade

Brad Sherman
Ranking Member, Subcommittee on Asia
and the Pacific

Over the past two years, as other social media companies worked to remove terrorist content, Telegram has become increasingly widely used by terrorists.

Following the recent news about Telegram's ICO for its cryptocurrency, MEMRI is again meeting with the same offices, as well as with new leaders on the Hill, with recommendations to deal with the threat of a Telegram cryptocurrency.

What Can Be Done: A Set Of Recommendations For Forcing Telegram To Remove Jihadi Content

The following are several steps to effect the removal of jihadi content from Telegram, in order to make sure the platform does not become a fundraising base for terrorists:

Legislature: Relevant panels of Congress can schedule hearings and invite Durov to testify and answer questions, under oath, about Telegram and its policies and practices relating to terrorist-linked content discovered on the platform, and about ensuring that terrorist fundraising will not take place on it. If he refuses, Durov can be subpoenaed and/or Congress can seek ways to penalize his or his company›s investments and operations in the U.S. – including but not limited to designating him and/or his company as terrorist entities.

Google And Apple: Since Google's Play Store and Apple's iTunes App Store are the only outlets for downloading the Telegram app, both these companies are in a unique position to pressure Telegram to remove jihadi content. If safeguards against jihadi use of Telegram are not added, then they can be asked to remove the app from their stores. Durov has expressed concern about the possibility of Telegram's removal from the App Store, and is vulnerable to threats in this regard. In February 2018, Telegram was briefly removed from the App Store after it was discovered that child pornography was being disseminated through its platform.[21]

> **Pavel Durov** ✓
> @durov
> [Follow ▾]
>
> Replying to @vondylan_ @telegram
>
> We were alerted by Apple that inappropriate content was made available to our users and both apps were taken off the App Store. Once we have protections in place we expect the apps to be back on the App Store.
>
> 1:07 AM - 1 Feb 2018 from Zurich, Switzerland
>
> 253 Retweets 337 Likes
>
> ♡ 81   ⟲ 253   ♡ 337

Telegram Servers And Presence In The U.S.: While not much is known about Telegram's presence in the U.S., the company appears to operate at least partly within its borders. On October 13, 2016, Telegram tweeted an announcement stating: "Issues in North and Latin America: cooling system died in one data center, massive overheating. Data safe, working with DC staff to fix." Telegram should be asked to provide details on the locations of its servers in the U.S.

> **Telegram Messenger** ✓
> @telegram
>
> @aepetrilli Telegram servers for users from Mexico are located in San Francisco.
>
> 2/7/14, 3:14 AM
>
> 1 RETWEET
>
> Provided by
> MEMRI JTTM

The Coming Storm: Terrorists Using Cryptocurrency

Additionally, Telegram states on its website that its servers are "spread worldwide for security and speed." Since Telegram operates within the U.S., the U.S. government may have legal and regulatory remedies for forcing Telegram to remove terrorist content, including fines or cessation of operations.

According to an October 13, 2016 Telegram tweet, the company has "DC staff" and its servers are "spread worldwide."

Previously Telegram Had No Go-To Contact Address; Now Their Lawyers Have Been Identified – And Should Be Questioned About Their Client: According to reports, the international law firm Skadden, Arps, Slate, Meagher, & Flom represents Telegram.[22] The firm could be notified of this issue, and asked to explain Telegram's policies and practices related to terrorist content on its platform.

Securities and Exchange Commission (SEC) Should Subpoena Durov And Telegram Like They Do With Other Cryptocurrencies: Earlier this year, the SEC sent a series of subpoenas on coin offerings; Telegram could be included in them. In February 2018, Telegram submitted to the SEC a filing signed by Durov stating that money had been raised "for the development of the TON Blockchain, the development and maintenance of Telegram Messenger, and other purposes." The security agreement is described as "purchase agreements for cryptocurrency."[23]

(Image: Sec.gov/Archives/edgar/data/1729650/000095017218000030/xslFormDX01/primary_doc.xml)

The SEC Document Listed Its Jurisdiction Of The Cryptocurrency As The British Virgin Islands: The island has a Financial Investigation Agency, which according to its website contains a list of acts and amendments pertaining to financial criminal activity, including several that deal with financial criminal acts including financial actions.[24] In 2017, the British Virgin Islands (BVI) amended its Anti-Money Laundering and Terrorist Financing amendment code of practice. According to the Department of State's Country Reports on Terrorism, the BVI is a U.K. overseas territory. The economy is dependent on tourism and the offshore financial sector. BVI authorities work with regional and U.S. law enforcement agencies to help mitigate the threats.[25] The BVI should be notified about issues of terrorism financing on a cryptocurrency connected with its territory.

Action By The U.S. Department Of Justice: The Department of Justice can designate organizations and individuals linked to terrorism. It can investigate, designate and/or sanction Telegram, Durov, and other key Telegram leaders, considering that terrorist entities now fundraise on Telegram and likely will continue to do so with cryptocurrency. Authority to name Specially Designated Nationals derives, in part, from 18 U.S.C. 2339B, the Antiterrorism and Effective Death Penalty Act of 1996 (AEDPA), which gave the Secretary of State authority to designate Foreign Terrorist Organizations (FTO) whose terrorist activity threatens the security of U.S. nationals or the national defense, foreign relations or economic interests of the U.S. The AEDPA made it unlawful, within the U.S. or for any person who is subject to the jurisdiction of the U.S. anywhere, to knowingly provide material support to a FTO that has been designated by the Secretary of State.[26]

Action By The U.S. Department Of Treasury: The Treasury Department can add Telegram and/or Durov to its list of Specially Designated Nationals (SDNs) because of it hosts terrorist content and facilitates terrorist fundraising – especially as it adds cryptocurrency and is poised to become a major conduit for terrorism finance. The Treasury Department's Office of Foreign Asset Control (OFAC) lists individuals, groups, and entities, including terrorists, designated under programs that are not country-specific. Collectively, such individuals and companies are called SDNs; their assets are blocked and U.S. persons are generally prohibited from dealing with them.[27] Treasury may also impose sanctions on persons determined to be responsible for or complicit in activities leading to specific harms caused by significant malicious cyber-enabled activities. Persons designated under this authority are added to OFAC's list of Specially Designated Nationals and Blocked Persons (SDN List). This authority, as amended, is intended to address situations where, for jurisdictional or other issues, certain significant malicious cyber actors may be beyond the reach of other authorities available to the U.S. government.[28] OFAC acts under Presidential emergency powers and authorities granted by specific legislation, but the sanctions it imposes also are based upon and in accordance with United Nations and other international mandates, are multilateral in scope, and involve close cooperation with U.S. allies.

International Community Activity: One of Telegram's largest population of users is located in Indonesia. In July 2017, Telegram agreed to block terrorist-related content in

Indonesia after the Indonesian government threatened to block the service over fears it was enabling terrorist communication. In response to the Indonesian threat, Durov said Telegram would remove ISIS-related channels flagged by the government and develop better systems for blocking such content in the future. On his Telegram channel, Durov stated: "It turns out that the officials of the Ministry recently emailed us a list of public channels with terrorism-related content on Telegram, and our team was unable to quickly process them. Unfortunately, I was unaware of these requests, which caused this miscommunication with the Ministry." Durov traveled to Indonesia, met with government officials there, and promised to remove terrorist content from the platform. This created a precedent for pressuring Telegram to act responsibly.

Pavel Durov ✔
@durov

Follow ⌄

In Jakarta, meeting with local teams brainstorming ways how to eradicate ISIS propaganda more efficiently.

11:50 PM - 31 Jul 2017 from Gambir, Indonesia

488 Retweets **891** Likes

💬 112   ⟲ 488   ♡ 891

Given Telegram's and Durov's worldwide activity, the international community and bodies involved with fighting terrorist fundraising need to get involved and demand that Durov account for Telegram's links to terrorism – if necessary, via subpoena, including the U.N., E.U., and in Dubai, where Telegram says it is currently located.

The United Nations Counter-Terrorism Committee Executive Directorate (CTED) works closely with other U.N. and external entities, such as the Financial Action Task Force (FATF) on steps to counter terrorism finance. Within the framework of the Counter-Terrorism Implementation Task Force (CTITF), CTED is a member of the Working Group on Countering the Financing of Terrorism.[29] CTED should have Durov testify under oath to explain what he is doing to remove terrorist content on Telegram and what his plans are to ensure that his platform and its cryptocurrency aren't used for terrorist fundraising. If he fails to do this, sanctions should be put in place.

Al-Qaeda Affiliate Hay'at Tahrir Al-Sham (HTS): Promoting And Soliciting Donations In Bitcoin; HTS Sheikh Declares It "Permissible" For Giving To Charity

«البتكوين» عملة اقتصاد المستقبل

Provided by MEMRI JTTM

أمـنـك التقني

الفريق التقني

اقتصرت العملة على مر الزمان على الذهب والفضة لما لهما من قيمة عالية وثبات في الظروف البيئة المختلفة، إلى أن تحولت للعملة الورقية ذات الغطاء الذهبي. ثم اقتصرت بعد «صدمة نيكسون» على العملة الورقية فقط دون غطاء الذهب، إلا أنها ظلت تتمتع بالقوة الشرائية اللازمة لجعلها «عملة» في النظام العالمي الاقتصادي القائم. أما في عصرنا الرقمي المعاصر فقد تغيرت المفاهيم وظهرت العملة الرقمية أو ما يسمى «البتكوين».

يمكننا القول أن «البتكوين» هو عملة إلكترونية يتم تداولها عبر الإنترنت، بدأ الحديث عنها لأول مرة عام 1998 وكانت الفكرة حول إنشاء شكل جديد من المال لا تتحكم فيه أي سلطة مركزية، وقد طبقت الفكرة عام 2009 لأول مرة ثم أخذت في الانتشار والتداول والتطوير حتى اعترفت بها العديد من دول العالم لم يكن آخرها الولايات المتحدة الأمريكية.

أما عن ماهية عمل «البتكوين» فيمكن التفكير فيه كبرنامج تقوم بتحميله لينشئ لك محفظة شخصية ذات عنوان خاص بها، ويتم تحويل العملات عبر العناوين بين طرفي البيع بدون وسيط وبشكل مشفر، ويعتمد هذا البرنامج على كود برمجي مفتوح المصدر بمعنى أنه يمكن لأي مطور أو مبرمج الاطلاع على هذا الكود وتطويره.

يزداد عدد مستخدمي «البتكوين» ويتسع الاعتماد عليه يوميا، حيث دخل في العديد من المعاملات مثل العقارات والمطاعم والمؤسسات العامة وخدمات الإنترنت، وذلك أن تعلم أن سعر البتكوين الواحد كان أقل من دولار واحد في عام 2009. إلى أن وصل في ديسمبر 2017 لأكثر من 17000$، وما زالت قيمته تتأرجح حسب العرض والطلب.

يتمتع البتكوين بالعديد من المزايا التي أكسبته الثقة والقوة الشرائية الحالية، ومنها:
- حرية الدفع: فمن الممكن إرسال واستقبال أي مبلغ بشكل لحظي ومشفر عبر العالم دون حدود أو قيود.

- رسوم قليلة جدا: تتم مدفوعات البتكوين إما بدون رسوم على الإطلاق أو برسوم قليلة جدا حيث يمكن للمستخدم تضمين رسوم نقل مع مدفوعاتهم للحصول على أولوية تنفيذ للمعاملة.

- أكثر أمنا: معاملات البتكوين آمنة وغير قابلة للعكس ولا تحتوي على أي معلومات خاصة عن صاحبها -بخلاف الحسابات البنكية وبطاقات الائتمان- وهذا يحمي المستخدم من الخسارة الناشئة عن محاولات الاحتيال المعروفة.

تعتبر العملة الرقمية بداية فكرة ثورية اجتاحت العالم لتكون أحد أهم الاختراعات في العقد الأخير وتهدد هذه التقنية النظام البنكي العالمي. فهل سنرى الزمن الذي سيحل فيه البتكوين بديلا عن العملات الورقية في العالم؟!

ebaa.news   شعبان 1440هـ

April 2019: Al-Qaeda-Affiliated Hay'at Tahrir Al-Sham (HTS) Media Arm Article Calls Bitcoin "Future Currency Of Economy"

On April 18, 2019, Ebaa', the news agency of Hay'at Tahrir Al-Sham (HTS), published an article about bitcoin, which it called "the future currency of economy." Following a history of the use of gold, silver, currencies, and banknotes, it went on to explain that bitcoin was "an application which, once installed, creates a personal wallet linked to a specific address and the fund will be encrypted and directly transferred between buyers and sellers."

It stated that bitcoin's use was growing worldwide and that it was now accepted in real-estate transactions, at restaurants, and by Internet service providers, and that it offers a flexible payment option, low fees, and high security. Bitcoin is, it added, a "revolutionary idea" that challenges the current international banking system, and queried whether it would replace banknotes worldwide.[30]

April 2019: Russian-Speaking Journalist Covering Hay'at Tahrir Al-Sham (HTS) In Syria Solicits Donations In Bitcoin On Telegram, Twitter, Other Platforms

Russian-speaking journalist Faruq Shami, who operates in rebel-held areas of Idlib, Aleppo, and northern Hama, in Syria, has interviewed members of Malhama Tactical (MT), Hay'at Tahrir Al-Sham (HTS), Tawhid Wal Jihad, and other jihadi groups. He often interviews Russian-speaking jihadis in Syria, from Russia, Uzbekistan, and Tajikistan. Shami frequently posts content critical of the Islamic State (ISIS), and is often embedded with

HTS, and links to other websites posting jihadi content. Shami operates on Facebook, Instagram, YouTube, Twitter, Telegram, and a personal blog, Muhajeer.com. In April 2019, he announced on Telegram that he was now accepting donations in bitcoin, and provided the bitcoin address and a QR code. He also has a pinned tweet on his Twitter account providing the same bitcoin information given on his Telegram channel.[31]

July 2019: Syria-Based Hay'at Tahrir Al-Sham (HTS) Jurist Abu Al-Fath Al-Farghali Declares Bitcoin Permissible For Giving To Charity

On July 13, 2019, Hay'at Tahrir Al-Sham jurist Sheikh Abu Al-Fath Al-Farghali posted a video on his Telegram channel, which has over 10,000 members, declaring bitcoin "permissible" according to shari'a, and permitting the giving of zakat (charity) using it.

In the 26-minute video, Al-Farghali, an Egyptian-born cleric who spent time in Sudan and Egypt before moving to Syria, began by relating the history of money and explaining how bitcoin developed. He said that bitcoin is a digital currency with no physical form and that it is extremely difficult to counterfeit or duplicate it, adding: "In my opinion, bitcoin, as it is today, is safer than paper currencies. True, the price [of bitcoin] is fluctuating, but this is also the case with regard to paper currencies. Some paper currencies have a certain degree of stability, but [the value of] paper currencies tends to fluctuate, in general."

Like gold, he said, bitcoin has a limited and finite amount that is 21 million that can be mined in total, and that this helps prevent the creation of inflation in the market. It is recognized as a legal currency in many countries, he said, including the U.S. and countries in Europe, as well as some Asian and African countries.

Addressing the current price volatility of bitcoin, Al-Farghali attributed it to speculation and to the fact that there are many investors despite bitcoin's limited supply. Some people, he added, have lost trust in bitcoin because of its, instability just like with any other currency, and went on to underline that a key advantage of bitcoin is that no country controls it.

He went on to say that it is permissible, as stated by several Muslim scholars, to use bitcoin just like other banknotes with no gold reserve behind them, as people have trusted and started dealing with them. Indicating that bitcoin is resistant to fraud and forgery when compared to banknotes, he noted that there are still risks because it is an emerging market and that people might lose trust in Bitcoin and stop using it, which could reduce its value to zero, just like any other banknote. "In my opinion," he said, "bitcoin in its current standing is more secure than banknotes with regard to their credit value."

He added: "Thus, as I said, it is permissible to use [virtual] currencies, which are closer to gold and silver in terms of their credit value. They should be used to pay zakat [charity] if you have a certain amount that is equal to a [similar] amount in gold or silver."[32]

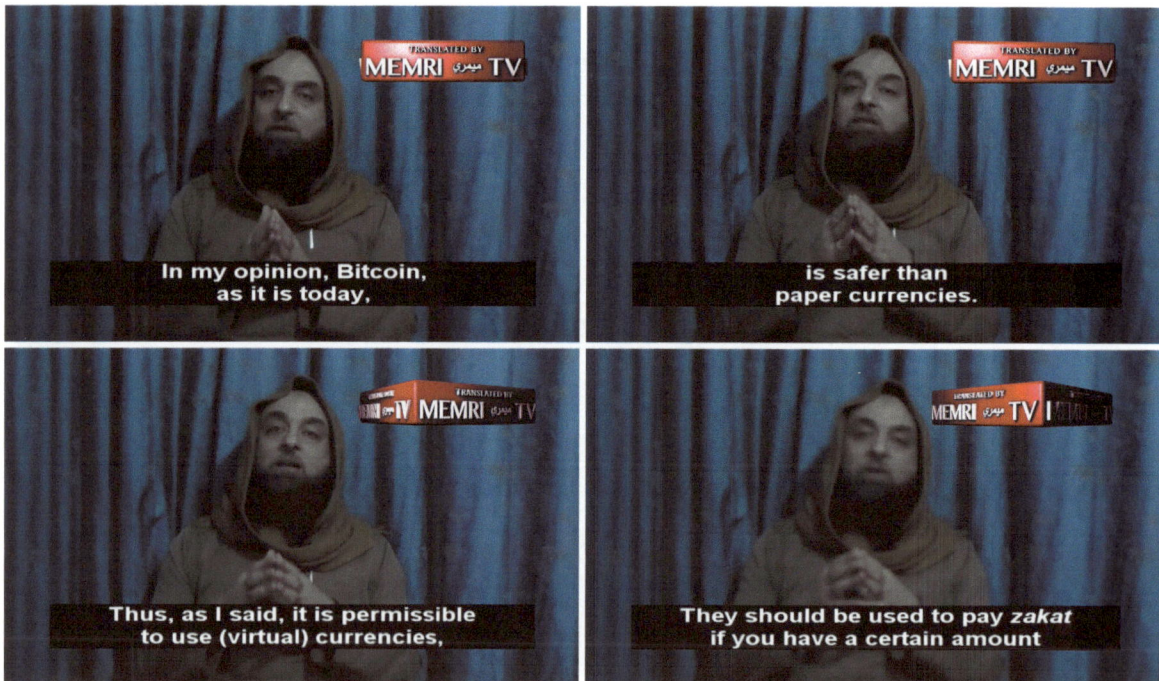

The Coming Storm: Terrorists Using Cryptocurrency

SadaqaCoins – Jihadi Crowdfunding Platform Using Cryptocurrencies – Including Bitcoin, Monero, And Ethereum; Encourages Hacking, Use Of Ransomware To Raise Funds; Posts Internet Security Tutorials; Active On Social Media

SadaqaCoins is a crowdfunding platform created to act as a secure connection between those who want to fund jihad and jihadis who need funding. The group invites those who need funding to set up projects on their website, and then raises funds for them, encouraging donors to use the Bitcoin, Monero, or Ethereum cryptocurrencies to donate. One project on the website is "Project: We Hunt," which seeks to raise funds to equip trained snipers.[33]

Promoting itself as one of the most "anonymous and secure" ways to finance the mujahideen, the group has email addresses on encrypted email services ProtonMail and Tutanota, among others, as well as a Twitter account and a Telegram channel. As this section will review, the group encourages followers to "encrypt databases, NAS and servers to blackmail owners to pay you a ransom" to raise funds but cautions that the hacker "ensure that no Muslims are mistakenly hit or the targets during your raids." It also calls on them to "find a suitable way to wage Jihad with [their] wealth" and to use cryptocurrencies such as Monero because they are "non-traceable."[34] SadaqaCoins also gives information on how to use the dark web and applications such as Orbot and Orfox, which make the user more anonymous online.

This section reviews SadaqaCoins and how it works, the ways supporters can contact the group and donate to it, how those seeking donations can set up new projects on the platform, active projects, and social media accounts, among other topics.

About SadaqaCoins

A page on the SadaqaCoins website explains what the group does and how its projects work, and answers the question "Why finance the mujahidin?" According to the group's site, SadaqaCoins enables "brothers and sisters to support the jihad financially along with complete anonymity." The group serves "as an intermediary between the financier and the Mujahidin working on these projects, whilst concealing and insulating the identity of both parties." Donors can contribute to the different projects listed on the group's website.

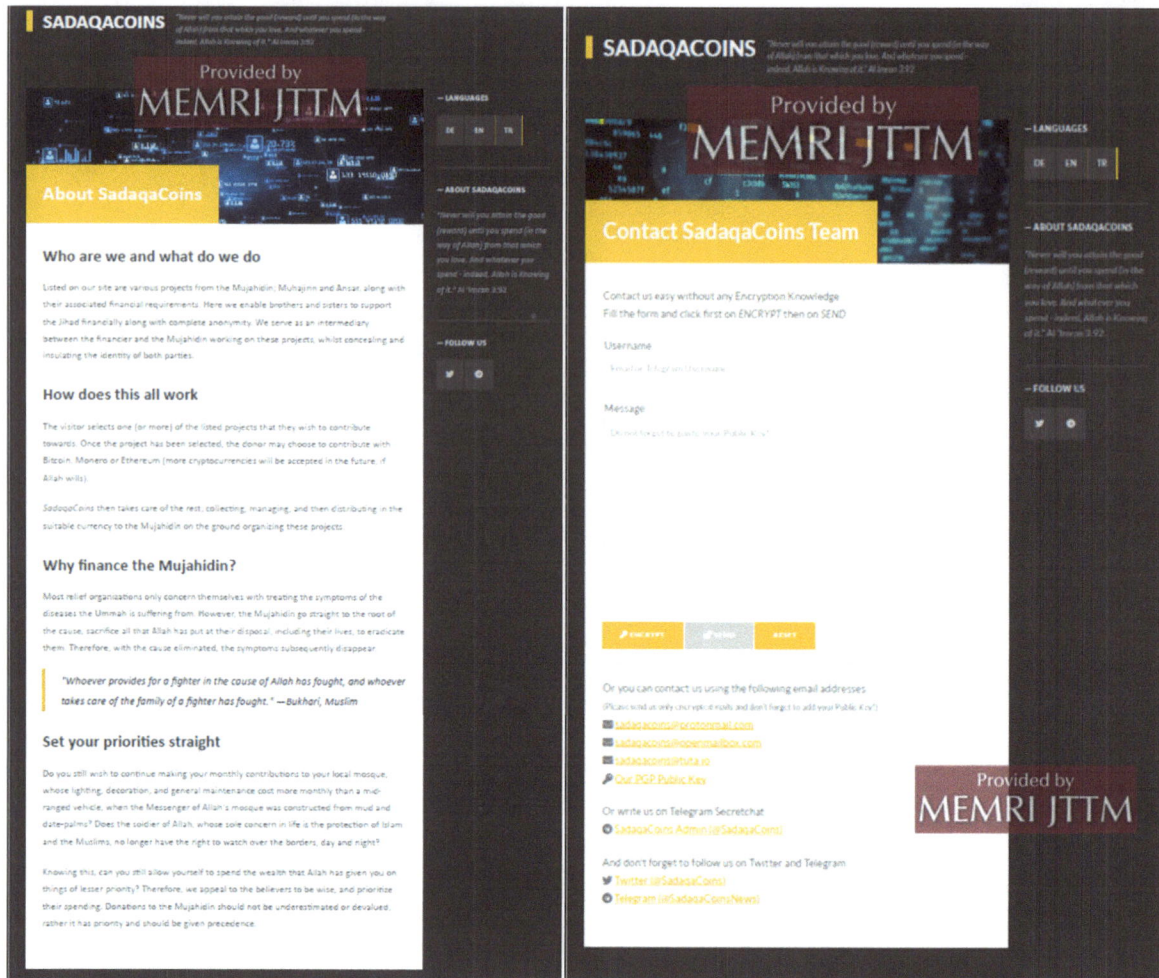

The website provides a form by which people can contact the group. The page asks the user to give a username, write a message, and then click "Encrypt" and "Send." The page gives SadaqaCoins' social media information, including its pages on Telegram (@sadaqacoins and @sadaqacoinsnews), and Twitter (@sadaqacoins). SadaqaCoins lists three different email addresses, including accounts with encrypted email services ProtonMail and Tutanota: sadaqacoins@protonmail.com, sadaqacoins@openmailbox.com, and sadaqacoins@tuta.io .

Frequently Asked Questions About Donating And Cryptocurrency

SadaqaCoins provides a page for Frequently Asked Questions. The first question it posts is: "Is SadaqaCoins a cryptocurrency?" The response reads: "No, SadaqaCoins is not a cryptocurrency, it is merely the name of a crowdfunding platform that acts as an interface between funders and project organizers." The second question addressed is: "Which secure

The Coming Storm: Terrorists Using Cryptocurrency

payment methods does SadaqaCoins prefer?" The response is: "We prefer Bitcoins, as it is the most well-known, rapidly growing cryptocurrency which is easy to access. However, we prefer Monero because it is 100% anonymous, non-traceable and is safer than Bitcoin..." A third question asks if donations could be made using PayPal. The response is that Sadaqa-Coins recommends only using cryptocurrency, but that "Coupons" could be accepted in special circumstances.

Frequently Asked Questions

Got a question? Send us a message and we will answer it, if Allah wills.

Is SadaqaCoins a cryptocurrency?

No, SadaqaCoins is not a cryptocurrency, it is merely the name of a crowdfunding platform that acts as an interface between funders and project organizers.

Which secure payment methods does SadaqaCoins prefer?

We recommend Bitcoins, as it is the most well-known, rapidly growing cryptocurrency which is easy to access. However, we prefer Monero because it is 100% anonymous, non-traceable and is safer than Bitcoin and Ethereum transactions.

If Ethereum or Bitcoins are not secure enough for you, you can exchange them for Monero and then send it to us. You can also use mixers to cover your tracks. However, there is a higher risk of fraud with mixers. (Recommend exchanger xmr.to and shapeshift.io)

I don't have Cryptocurrency. Can I also donate with something like Paypal?

For Security reasons we only recommend using Cryptocurrency. If you are unable to obtain Cryptocurruncies, you may contact us, and in special circumstances we can possibly accept Coupons and other alternative means of donations

The page gives the groups addresses for Bitcoin[35] and Monero[36] so that supporters can donate. The group writes: "If you prefer a private address, please contact us." In response to the notion that "Bitcoin and Cryptocurrencies are Haram [forbidden]!" SadaqaCoins says: "According to the principle of *fiqh* [Islamic jurisprudence], 'that whatever an obligatory acts requires in order to perform it, is itself obligatory' it becomes obligatory to find a subtle way to wage Jihad with the wealth and since we recognize that it is not possible for every believer to personally deliver their wealth to the fronts of jihad, this is at our present, our next best method."

But Bitcoin and Cryptocurrencies are Haram!

Firstly, it is true that there are a number of different opinions regarding the issue. However, in some situations such as ours, it is not always easy to obtain fiat currency without putting all parties at risk.

Nevertheless, Jihad with the wealth is as much, if not more of an obligation upon the believer than Jihad with the self (i e fighting) as is evident in numerous places throughout the Qur'an and Sunnah, thus there is no question of the believers abandoning it

Therefore, according to the principle of Fiqh, 'that whatever an obligatory acts requires in order to perform it, is itself obligatory' it becomes obligatory to find a suitable way to wage Jihad with the wealth, and since we recognise that it is not possible for every believer to personally deliver their wealth to the fronts of Jihad, this is at present, our next best method

While, if in the future more suitable ways of waging Jihad with the wealth become available to us, then it will become an obligation upon us to adopt them instead of current methods, and Allah knows best

Yet, if you are still in doubt, then by all means, consult students and scholars known for their truthfulness and love for Jihad. However, if you are convinced of the impermissability of Cryptocurrencies like Bitcoin, then we respect your opinion.

Is it necessary to support a Project? Can I not just donate directly to SadaqaCoins?

Of course you can support SadaqaCoins directly Use the following public addresses

🅱 Bitcoin Donation: 14gymFijxkFzbxbacbP9ioGndsqHRuJJTc

Ⓜ Monero Donation 42eTgdYrtCvBWxHhnLxZY3MKwouiJLwaCLMMyD8YhwDvfp tQkLCSbJXYZCcVk6EB3gNHVvXNuDgqq24dk9mxqHoBJY8L7PV

If you prefer a private address, please contact us

Can I also promote my own Project(s) on SadaqaCoins?

Yes, as long as the project on the ground is reasonable, we will consider it for promotion aslong as certain conditions are met Contact us in order to discuss the details of your proposed project

The Coming Storm: Terrorists Using Cryptocurrency

Four Digital Ways To Support SadaqaCoins

SadaqaCoins lists four ways to support the group: "advertising," "buy cryptocurrency," "mine cryptocurrency," and "hustle cryptocurrency." The first way to support SadaqaCoins, advertising, involves spreading awareness of the group's projects via social media. The second way is through buying cryptocurrency locally rather than using online services that can be linked back to the buyer.

The third way to support the group described on its website is to download the programs necessary and mining cryptocurrencies such as Monero or Ethereum. Lastly, SadaqaCoins suggests "hustling" cryptocurrency by hacking Bitcoin and "Ethereum Miner farms" and redirecting transactions. SadaqaCoins says: "Encrypt databases, NAS and servers to blackmail owners to pay you a ransom. Take precautions to ensure that no Muslims are mistakenly hit or the targets during your raids. Since these computer attacks happen through a kind of 'violence' and one fears an arrest, this category falls under *ghanima* (war booty). You can pay a part of it to SadaqaCoins."

Donation-Seekers Can "Submit A New Project" To Raise Funds

To submit a new project to SadaqaCoins, the organizer must enter his or her name, contact details, the name of the project, a description of the project, and target goals. The form asks for credentials for potential backers, financial goals, if the organizer wishes for the project to be public or private, and the name of the group to which the organizer belongs.

Submit a new Project

— LANGUAGES

DE EN TR

— ABOUT SADAQACOINS

"Never will you attain the good (reward) until you spend (in the way of Allah) from that which you love. And whatever you spend - indeed, Allah is Knowing of it." Al 'Imran 3:92

— FOLLOW US

Please fill in the following fields carefully and in detail.
Fill out the form and first click *ENCRYPT* and then *SEND*.

Name of the Organizer:

Who is responsible?

Contact details of the Organizer:

Telegram Username or Email Address

Name of the project:

What should your project be called on SadaqaCoins?

Description of the project:

Describe the project in detail and provide all necessary information. The more you explain about your project and your goals in detail, the more likely it will be approved, and supported.

Target/goals you require for the project:

List any resources needed, and their associated expected costs.

Status of the project:
- ⦿ Public
- ◯ Private

Credentials for potential backers: (Fill in only if project status is private)

Username

Password

Financial goal: (Include currency)

$1000

Name of the group you belong to: (Fill only if you belong to a group)

Name of the group

Message:

If you wish to include additional information, or have any queries, you can do it here. If you have not provided a Telegram Username, then an Email Address is required, but do not forget to include a PGP key.

The Coming Storm: Terrorists Using Cryptocurrency

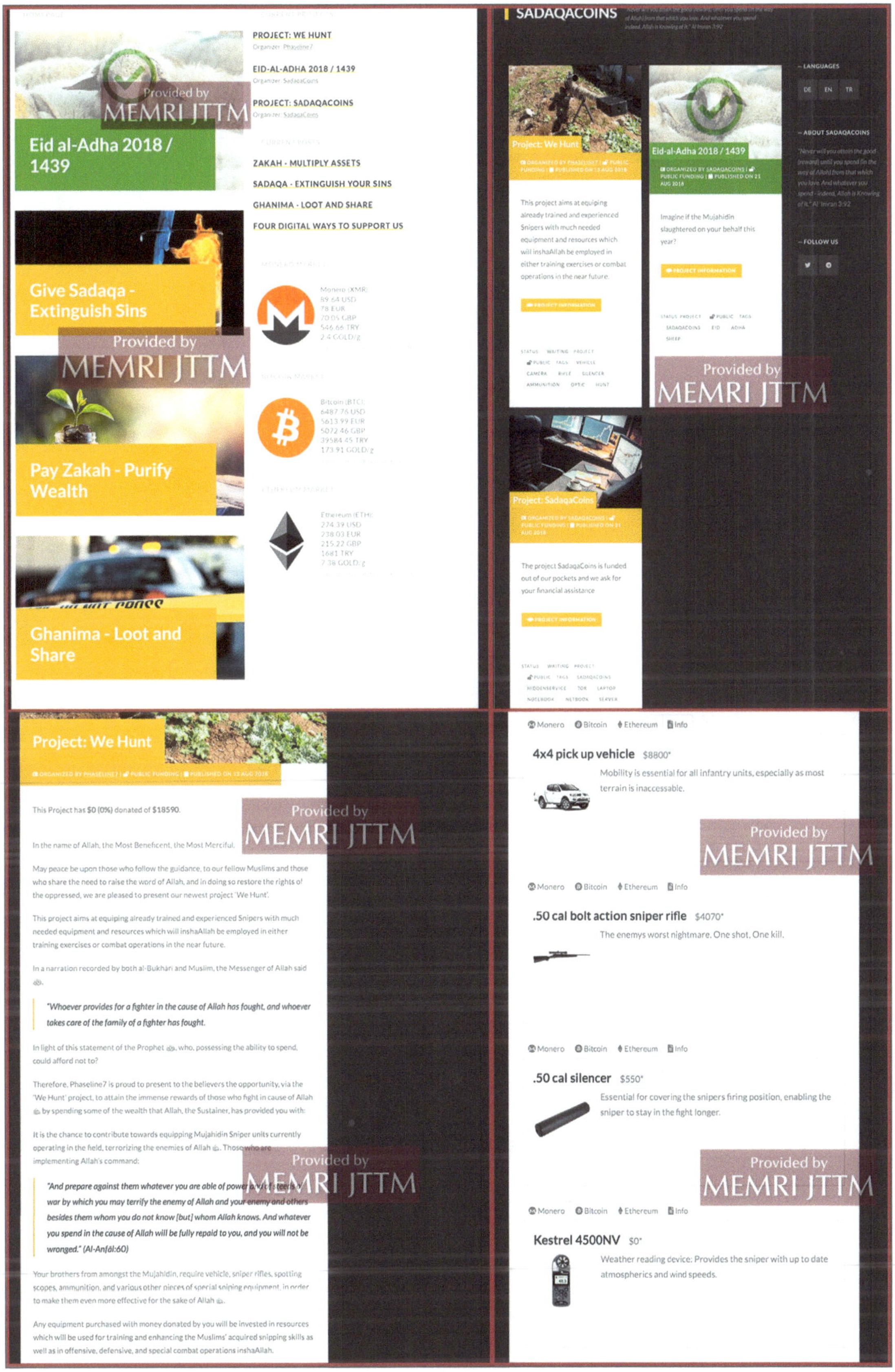

Left panel (top)

Provided by MEMRI JTTM

Eid al-Adha 2018 / 1439

PROJECT: WE HUNT
Organizer: Phaseline7

EID-AL-ADHA 2018 / 1439
Organizer: SadaqaCoins

PROJECT: SADAQACOINS
Organizer: SadaqaCoins

ZAKAH · MULTIPLY ASSETS

SADAQA · EXTINGUISH YOUR SINS

GHANIMA · LOOT AND SHARE

FOUR DIGITAL WAYS TO SUPPORT US

Give Sadaqa - Extinguish Sins

Provided by MEMRI JTTM

Pay Zakah - Purify Wealth

Ghanima - Loot and Share

Monero (XMR):
89.64 USD
78 EUR
70.05 GBP
546.66 TRY
2.4 GOLD/g

Bitcoin (BTC):
6487.76 USD
5613.99 EUR
5072.46 GBP
39584.45 TRY
173.91 GOLD/g

Ethereum (ETH):
274.39 USD
238.03 EUR
215.22 GBP
1681 TRY
7.36 GOLD/g

Right panel (top)

SADAQACOINS

Project: We Hunt
ORGANIZED BY PHASELINE7 | PUBLIC FUNDING | PUBLISHED ON 13 AUG 2018

This project aims at equiping already trained and experienced Snipers with much needed equipment and resources which will inshaAllah be employed in either training exercises or combat operations in the near future.

STATUS: WAITING PROJECT
PUBLIC TAGS VEHICLE
CAMERA RIFLE SILENCER
AMMUNITION OPTIC HUNT

Eid al-Adha 2018 / 1439
ORGANIZED BY SADAQACOINS | PUBLIC FUNDING | PUBLISHED ON 21 AUG 2018

Imagine if the Mujahidin slaughtered on your behalf this year?

STATUS PROJECT | PUBLIC TAGS
SADAQACOINS EID ADHA
SHEEP

Project: SadaqaCoins
ORGANIZED BY SADAQACOINS | PUBLIC FUNDING | PUBLISHED ON 31 AUG 2018

The project SadaqaCoins is funded out of our pockets and we ask for your financial assistance.

STATUS: WAITING PROJECT
PUBLIC TAGS SADAQACOINS
HIDDENSERVICE TOR LAPTOP
NOTEBOOK NETBOOK SERVER

LANGUAGES
DE EN TR

ABOUT SADAQACOINS
"Never will you attain the good [reward] until you spend [in the way of Allah] from that which you love. And whatever you spend - indeed, Allah is Knowing of it." Al 'Imran 3:92

FOLLOW US

Provided by MEMRI JTTM

Left panel (bottom)

Project: We Hunt
ORGANIZED BY PHASELINE7 | PUBLIC FUNDING | PUBLISHED ON 13 AUG 2018

This Project has $0 (0%) donated of $18590.

In the name of Allah, the Most Beneficent, the Most Merciful.

May peace be upon those who follow the guidance, to our fellow Muslims and those who share the need to raise the word of Allah, and in doing so restore the rights of the oppressed, we are pleased to present our newest project 'We Hunt'.

This project aims at equiping already trained and experienced Snipers with much needed equipment and resources which will inshaAllah be employed in either training exercises or combat operations in the near future.

In a narration recorded by both al-Bukhari and Muslim, the Messenger of Allah said ﷺ.

"Whoever provides for a fighter in the cause of Allah has fought, and whoever takes care of the family of a fighter has fought."

In light of this statement of the Prophet ﷺ, who, possessing the ability to spend, could afford not to?

Therefore, Phaseline7 is proud to present to the believers the opportunity, via the 'We Hunt' project, to attain the immense rewards of those who fight in cause of Allah ﷺ by spending some of the wealth that Allah, the Sustainer, has provided you with:

It is the chance to contribute towards equipping Mujahidin Sniper units currently operating in the field, terrorizing the enemies of Allah ﷺ. Those who are implementing Allah's command:

"And prepare against them whatever you are able of power and of steeds of war by which you may terrify the enemy of Allah and your enemy and others besides them whom you do not know [but] whom Allah knows. And whatever you spend in the cause of Allah will be fully repaid to you, and you will not be wronged." (Al-Anfāl:60)

Your brothers from amongst the Mujahidin, require vehicle, sniper rifles, spotting scopes, ammunition, and various other pieces of special sniping equipment, in order to make them even more effective for the sake of Allah ﷺ.

Any equipment purchased with money donated by you will be invested in resources which will be used for training and enhancing the Muslims' acquired snipping skills as well as in offensive, defensive, and special combat operations inshaAllah.

Provided by MEMRI JTTM

Right panel (bottom)

Monero Bitcoin Ethereum Info

4x4 pick up vehicle $8800*

Mobility is essential for all infantry units, especially as most terrain is inaccessable.

Provided by MEMRI JTTM

Monero Bitcoin Ethereum Info

.50 cal bolt action sniper rifle $4070*

The enemys worst nightmare. One shot, One kill.

Monero Bitcoin Ethereum Info

.50 cal silencer $550*

Essential for covering the snipers firing position, enabling the sniper to stay in the fight longer.

Provided by MEMRI JTTM

Monero Bitcoin Ethereum Info

Kestrel 4500NV $0*

Weather reading device: Provides the sniper with up to date atmospherics and wind speeds.

Stalinsky M E M R I ميمري

Current Posts And Projects: "Give Sadaqa – Extinguish Sins," "Project: We Hunt," And "Ghanima – Loot And Share"

SadaqaCoins gives the names of its projects as well as information on the Bitcoin, Ethereum, and Monero markets. The page includes currency exchange rates for each of the markets in U.S. dollars, Euros, Pounds Sterling, Turkish Lira, and gold.

One of the projects on the platform is "Project: We Hunt," which equips snipers and "Project: Sadaqacoins," which provides donations to help the SadaqaCoins team.

"Project: We Hunt" and "Project: Sadaqa-Coins," are explained in detail. "Project: We Hunt" "aims at equiping already trained and experienced Snipers with much needed equipment and resources which will inshaAllah be employed in either training exercises or combat operations in the near future." The project appears to have raised no funds toward its goal of $18,590. Sadaqa-Coins presents this project as an opportunity "to attain the immense rewards of those who fight in cause of Allah." Some of the items listed as necessary for "Project: We Hunt" are:

- "4x4 pick up vehicle" for $8,800
- ".50 cal bolt action sniper rifle" for $4,070
- ".50 cal silencer" for $550
- "Kestrel 4500NV" weather-reading device, for which no price is given
- Nikon camera for $550
- Rifle optics for $3,960
- Ammunition for $660.

Each item can be paid for via the Monero, Bitcoin, or Ethereum cryptocurrencies.

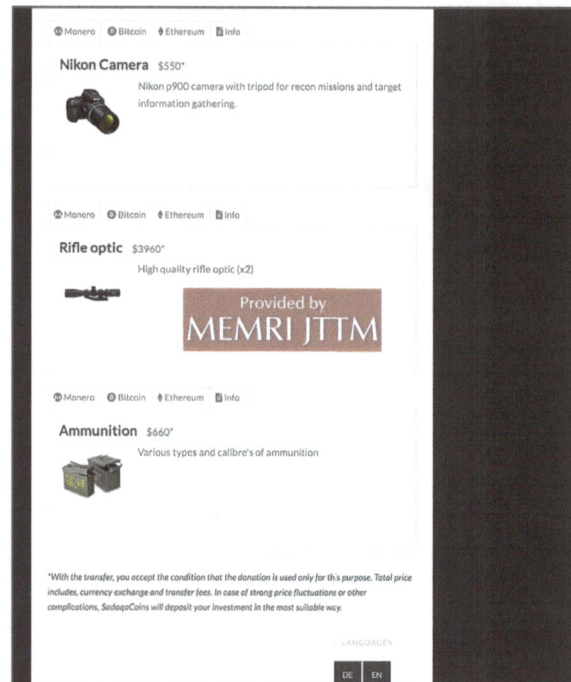

The donations for "Project: SadaqaCoins" go toward the expansion of SadaqaCoins. The group appeals to donors saying: "With your generous donations we can promote SadaqaCoins, so that together, it can grow into a revolutionary platform facilitating believers perform Jihad with their wealth in a secure and anonymous fashion." Donations to this project go toward funding a Tor HiddenService Server for $160, a laptop for $660, "hardware cold-wallet" for the secure storage of cryptocurrencies for $110, and translation work for $110. There is a note at the bottom of "Project: SadaqaCoins" that specifies: "With the transfer, you accept the condition that the donation is used only for this purpose. Total price includes, currency exchange and transfer fees. In case of strong price fluctuations or other complications, SadaqaCoins will deposit your investment in the most suitable way."

The Coming Storm: Terrorists Using Cryptocurrency

Project: SadaqaCoins

ORGANIZED BY SADAQACOINS | PUBLIC FUNDING | PUBLISHED ON

This Project has $0 (0%) donated of $600.

The SadaqaCoins Team volunteer for the sake of Allah, the Almighty, and we operate a strict 100% donations policy, which means that 100% of the money you donate will be used to fund your chosen project.

However, we require your financial assistance for a number of things: By the permission of Allah, in the future, we hope to be able to expand our service to include a number of languages such as Arabic, French, Spanish, and Russian.

This expansion, if Allah wills, will allow us to improve the quality and scope of the project. With your generous donations we can promote SadaqaCoins, so that together, it can grow into a revolutionary platform facilitating believers perform Jihad with their wealth in a secure and anonymous fashion.

Free doesn't mean free of cost

We do not intend to divert proceeds from donations to sources other than those of a given to the project you fund. Our team of volunteers can continue to do our work for free, however there are many fiscal factors that have to be addressed regarding SadaqaCoins maintenance.

Administration and maintaining the server, website design and managing the projects takes a lot of time and money. Especially since we, ourselves, are in a conflict area and the situation can turn from one moment to another.

Monero | Bitcoin | Ethereum | Info

Tor HiddenService Server $160*

Annual operating cost of the VPS / Dedicated Server

Provided by MEMRI JTTM

Monero | Bitcoin | Ethereum | Info

Intel i5-i7 CPU 8GB RAM Laptop $660*

The SadaqaCoins team require more laptops for efficient and effective work on SadaqaCoins. (2x)

Monero | Bitcoin | Ethereum | Info

Hardware Cold-Wallet $110*

Secure storage of cryptocurrencies

Provided by MEMRI JTTM

Monero | Bitcoin | Ethereum | Info

Translation Work $110*

Translation services for site content (per language)

With the transfer, you accept the condition that the donation is used only for this purpose. Total price includes, currency exchange and transfer fees. In case of strong price fluctuations or other complications, SadaqaCoins will deposit your investment in the most suitable way.

Ghanima - loot and share

PUBLISHED ON 11 MAR 2018

"And know that anything you obtain of war booty - then indeed, for Allah is one fifth of it and for the Messenger and for [his] near relatives and the orphans, the needy, and the [stranded] traveller, if you have believed in Allah and in that which We sent down to Our Servant on the day of criterion - the day when the two armies met. And Allah, over all things, is competent." 8:41

Ghanima (War booty) is the wealth, which is taken from the disbelievers through fighting. Imam Shafi'i said:

Ghanima is property that the Muslims seize from the disbelievers by means of overpowering them. Overpowering them includes using force openly or by deceiving them secretly since the Messenger of Allah, Peace be upon him, said that war is deception.

Therefore, according to Imam Shafi'i, money that taken from the disbelievers using clandestine methods should be considered Ghanima, even if the use of force is not involved. (Please refer: Four Digital Ways for Support)

"So consume what you have taken of war booty [as being] lawful and good, and fear Allah. Indeed, Allah is Forgiving and Merciful." 8:69

Did you loot something and you want to give a part of it? SadaqaCoins also accepts Cryptocurrency and uses it for different profects as required.

Bitcoin Donation: 34gyenFjxkFzbxbacbP9oGrdsqHBuLJ7c

Monero Donation:
42eTpdYrtCxBWxHhnLxZY3MKwouiLwuCLMM4yD8YhwQvfptOxLCSbJXY7CcVkAfJO8pNHVvXNuDgqn24dk9mxnHoBJY

TAGS GHANIMA DONATION CHARITY BITCOIN MONERO ETHEREUM CRYPTOCURRENCY JIHAD

LANGUAGES
DE EN TR

ABOUT SADAQACOINS

"Never will you attain the good [reward] until you spend [in the way of Allah] from that which you love. And whatever you spend - indeed, Allah is Knowing of it." Al Imran 3:92

FOLLOW US

Sadaqa - extinguish your sins

PUBLISHED ON 11 MAR 2018

Sadaqa is not an obligatory act of worship but beloved by Allah that brings you closer to Him, the Sustainer. It is also a way the believer purifies his wealth.

Allah says:

Provided by MEMRI JTTM

The example of those who spend their wealth in the way of Allah is like a seed [of grain] which grows seven spikes; in each spike is a hundred grains. And Allah multiplies [His reward] for whom He wills. And Allah is all-Encompassing and Knowing. 2:261

Those who spend their wealth [in Allah's way] by night and by day, secretly and publicly - they will have their reward with their Lord. And no fear will there be concerning them, nor will they grieve. 2:274

While the Prophet, peace be upon him, says, specifically about spending for Jihad:

"Whoever spends in the path of Allah, it would be multiplied for them 700 times."

So, if any one of you spends as little as $100 for Jihad, sincerely for the sake of Allah, he or she will be rewarded as if they spent $70,000 to raise the word of Allah to the uppermost!

Another advantage of giving sadaqa is the sins are expiated and harm is kept away.

"Charity (al-Sadaqa) extinguishes sin, just as water extinguishes fire." —At-Tirmidhi

You can decide if you want to invest into one of your personally selected projects or allow the SadaqaCoins Team to use it according to their discretion on what is most necessary.

Even if you only advertise our project, it will also count as Sadaqa, because the Prophet, peace be upon him, said:

"Enjoining all that is good is Sadaqa" —Bukhari

"So for this let the competitors compete." 83:26

Bitcoin Donation: 34gyenFjxkFzbxbacbP9oGrdsqHBuLJ7c

Monero Donation:
42eTpdYrtCxBWxHhnLxZY3MKwouiLwuCLMM4yD8YhwQvfptOxLCSbJXY7CcVkAfJO8pNHVvXNuDgqn24dk9mxnHoBJY

FOLLOW US

...reward until you spend [in the way of Allah] from that which you love. And whatever you spend - indeed, Allah is Knowing of it." Al Imran 3:92

FOLLOW US

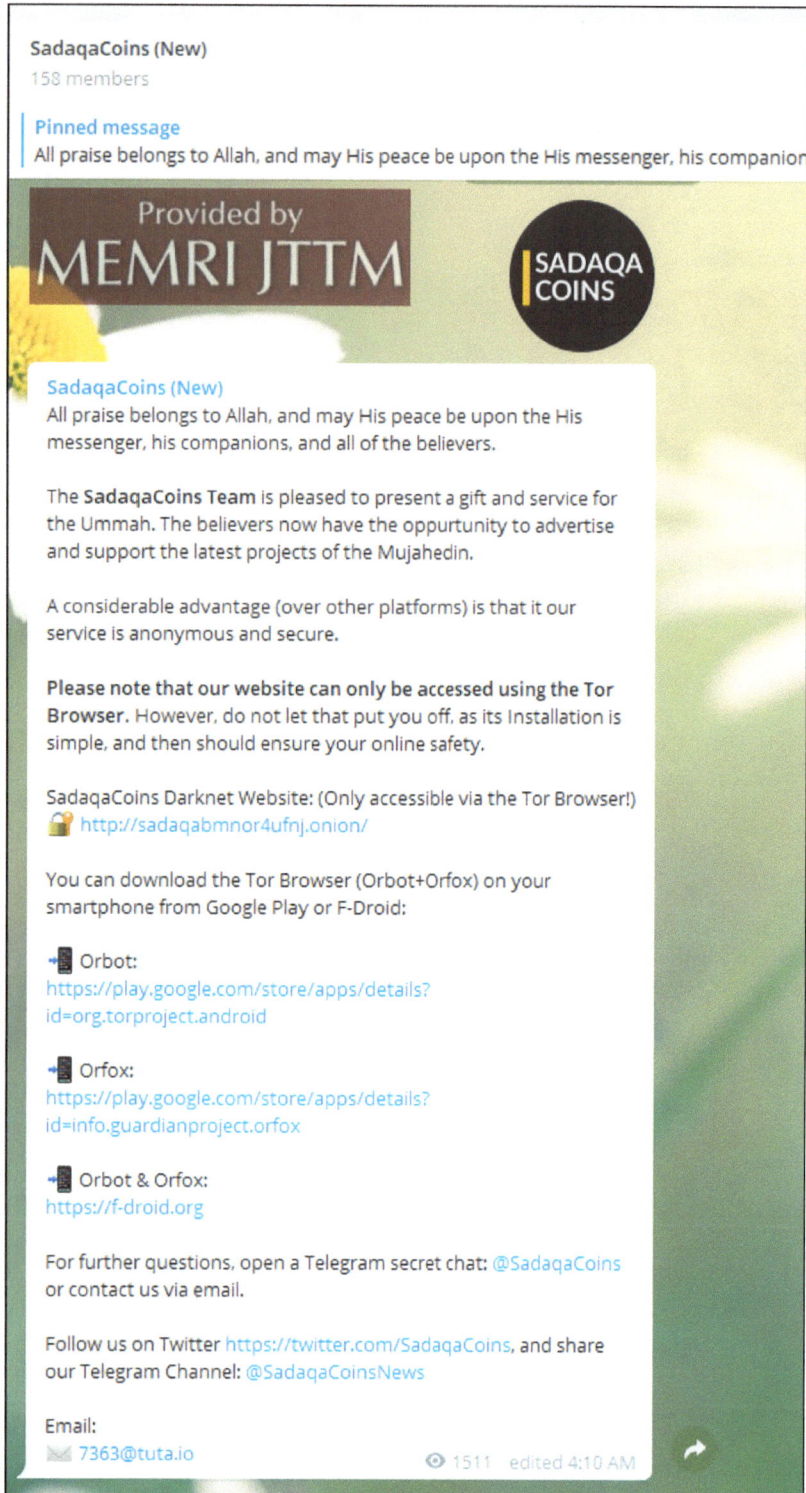

SadaqaCoins (New)
158 members

Pinned message
All praise belongs to Allah, and may His peace be upon the His messenger, his companion

Provided by
MEMRI JTTM

SADAQA COINS

SadaqaCoins (New)
All praise belongs to Allah, and may His peace be upon the His messenger, his companions, and all of the believers.

The **SadaqaCoins Team** is pleased to present a gift and service for the Ummah. The believers now have the oppurtunity to advertise and support the latest projects of the Mujahedin.

A considerable advantage (over other platforms) is that it our service is anonymous and secure.

Please note that our website can only be accessed using the Tor Browser. However, do not let that put you off, as its Installation is simple, and then should ensure your online safety.

SadaqaCoins Darknet Website: (Only accessible via the Tor Browser!)
🔒 http://sadaqabmnor4ufnj.onion/

You can download the Tor Browser (Orbot+Orfox) on your smartphone from Google Play or F-Droid:

📲 Orbot:
https://play.google.com/store/apps/details?id=org.torproject.android

📲 Orfox:
https://play.google.com/store/apps/details?id=info.guardianproject.orfox

📲 Orbot & Orfox:
https://f-droid.org

For further questions, open a Telegram secret chat: @SadaqaCoins or contact us via email.

Follow us on Twitter https://twitter.com/SadaqaCoins, and share our Telegram Channel: @SadaqaCoinsNews

Email:
✉ 7363@tuta.io

👁 1511   edited 4:10 AM

One of the posts from SadaqaCoins called "Ghanima – loot and share" discusses the Islamic concept of *ghanima* ("war booty"). The posts describes ghanima as the "property that the Muslims seize from disbelievers by means of overpowering them." The page reads: "Did you loot something and you want to give part of it? SadaqaCoins also accepts Cryptocurrency..." and then provides the group's Bitcoin and Monero addresses.

Another post, titled "Sadaqa – extinguish your sins," discusses the benefits of donating one's wealth and the divine rewards that donors will receive, citing hadith and verses from the Quran.

On September 25, 2018, SadaqaCoins posted instructions on Telegram on how to access the group's website using the Tor browser, providing links to the Orbot and Orfox mobile applications. These applications connect the user to the Tor network. The group also gave gave its Twitter address, @Sadaqacoins, and invite users to open a Telegram secret chat with @SadaqaCoins or contact them via email at 7363@tuta.io.

The Coming Storm: Terrorists Using Cryptocurrency

SadaqaCoins Releases Video of Militant Leader Encouraging Supporters to Donate Through Bitcoin

On January 3, 2019, SadaqaCoins disseminated a video titled "Jihadi Investment" via its Telegram channel showing Amir Muslim, the commander of the Junud Ash-Sham militant group in Syria, encouraging supporters to support the Jihad using bitcoin. The video, which isundated, was released in four different languages. In it, Amir Muslim says: "The brothers have prepared a secure channel through Bitcoin and everyone who wants to invest in this blessed cause can do it, if Allah wishes, and the brothers will explain where and how it can be invested. May Allah reward you with good."

Visible in the video is a watermark of SadaqaCoins' address on the dark web. The video was removed from the Telegram channel shortly after its release.

The SadaqaCoins Telegram channel, which has 150 members, provides tips on how to donate to jihad.[37]

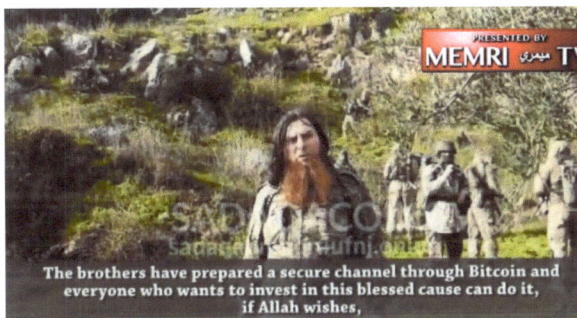

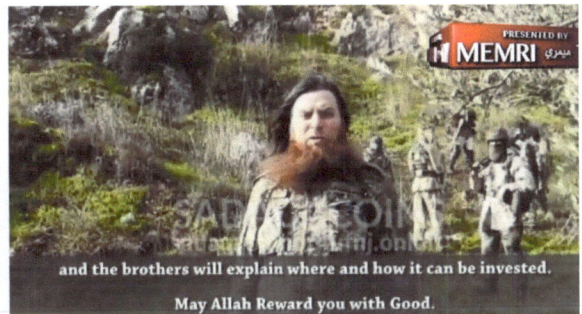

The brothers have prepared a secure channel through Bitcoin and everyone who wants to invest in this blessed cause can do it, if Allah wishes,

and the brothers will explain where and how it can be invested.
May Allah Reward you with Good.

SadaqaCoins Promotes Platform And Internet Security Tools, Connects With Other Jihadi Groups On Twitter

The SadaqaCoins Twitter account (@sadaqacoins), created in May 2018, has as of this writing 21 tweets and 29 followers, and is following 134 accounts. The description section reads: "The ummah needs you – support our troops..." The account gives a link to the group's website, Sadaqabmnor4ufnj.onion, which is accessible only using the Tor network, and often retweets other's posts about SadaqaCoins or articles about cryptocurrency. SadaqaCoins' Twitter account follows the jihadi news service On the Ground News, the jihadi fundraising group Al-Sadaqah, Documenting Oppression Against Muslims, the leader of jihadi military training group Malhama Tactical (MT), and others. The account has not posted since September 12, 2018.

In an August 17 tweet, SadaqaCoins advised followers to "Install and setup #Orbot."

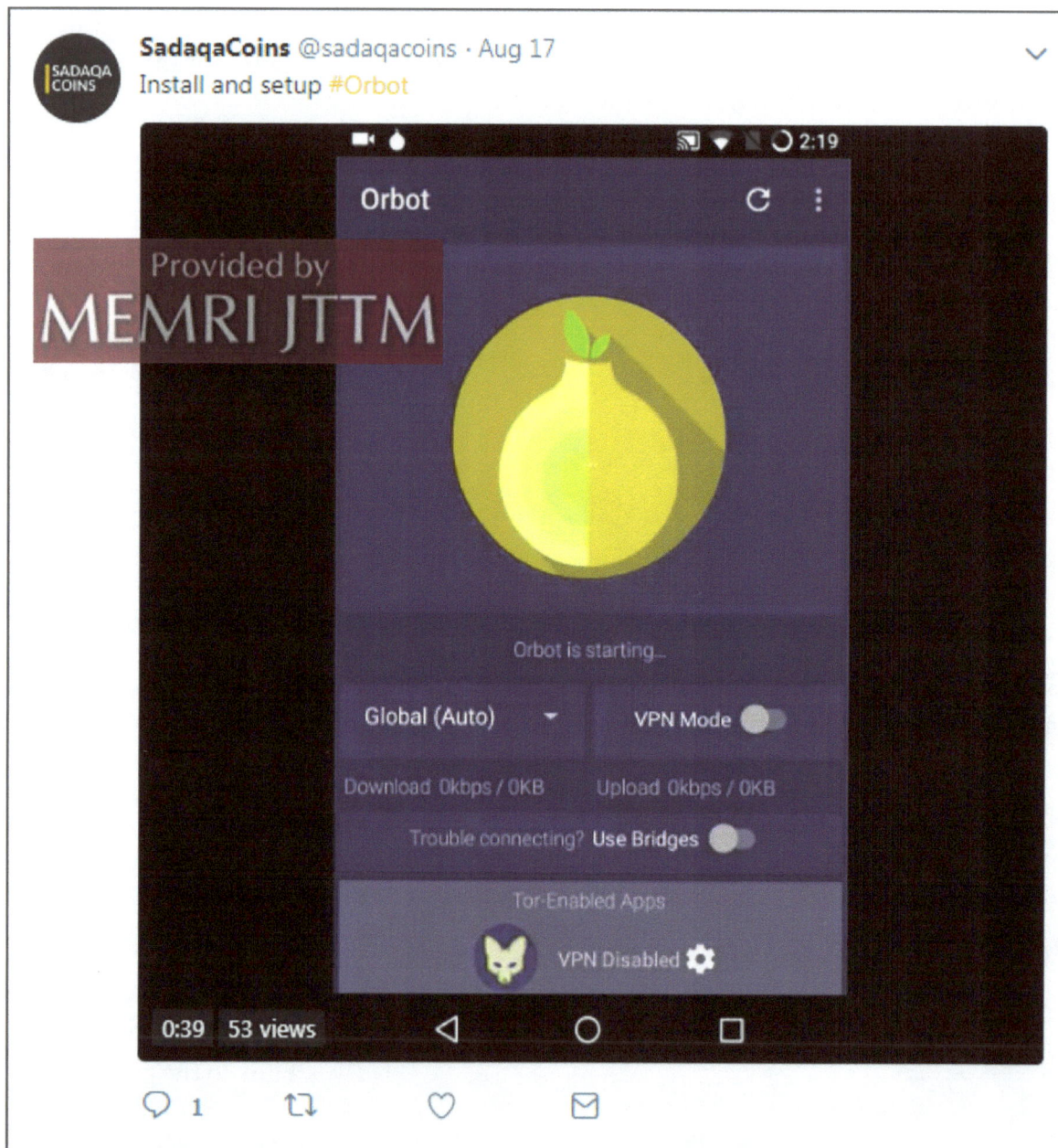

The Coming Storm: Terrorists Using Cryptocurrency

Also on August 17, SadaqaCoins tweeted: "Download #Fdroid, #Orbot and #Orfox" along with a photo showing how to download F-Droid, an installable catalogue of free and open source software applications for the Android platform.

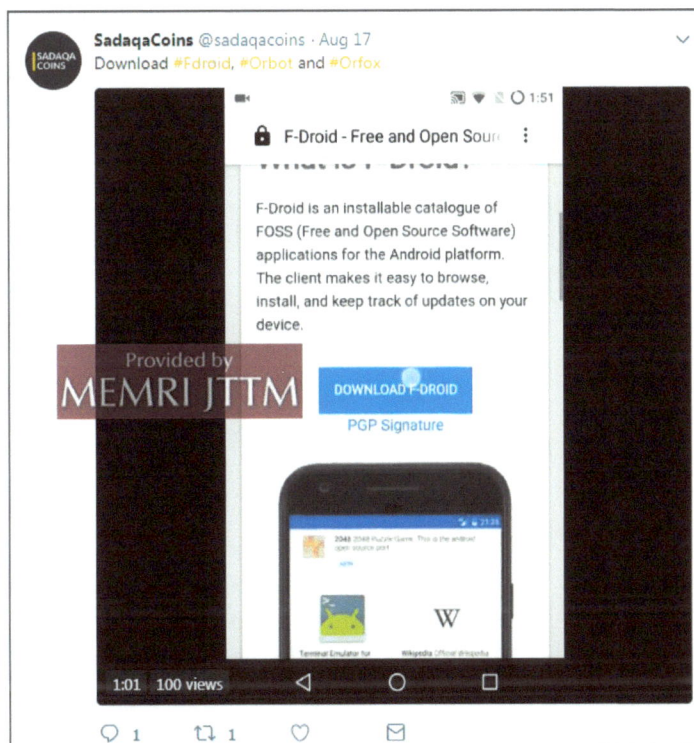

On August 23, SadaqaCoins posted a series of photos of live animals, followed by photos of cooked meat. This post coincided with Eid Al-Adha 2018.

SadaqaCoins Promotes Platform, Tor, Orbot, And Orfox On Telegram

SadaqaCoins opened a Telegram channel on September 25, 2018, that has 101 members. The channel's description section reads: "100% Donation Policy" and gives the link to the group's website.

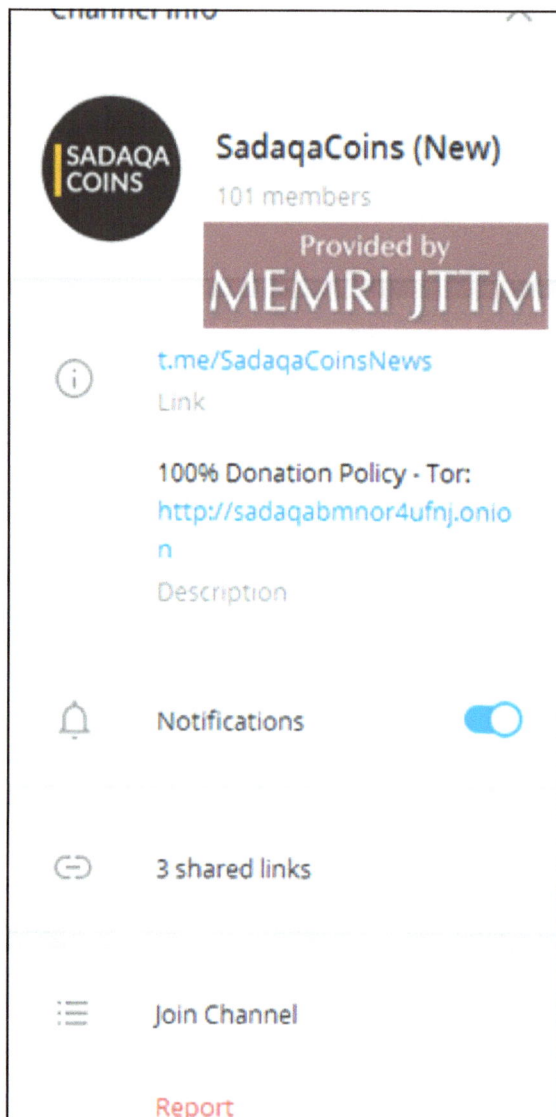

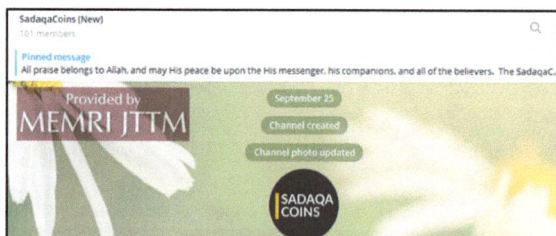

The Coming Storm: Terrorists Using Cryptocurrency

An early message on the group's channel describes SadaqaCoins: "The SadaqaCoins Team is pleased to present a gift and service for the ummah. The believers now have the opportunity to advertise and support the latest projects of the Mujahedin. A considerable advantage (over other platforms) is that our service is anonymous and secure. Please note that our website can only be accessed using the Tor Browser." The group gives links to download the Tor Browser, which allows for anonymous communication, as well as Orbot and Orfox. The message gives a link to the group's Twitter account and an address with the encrypted email service Tutanota: 7363@tuta.io.

SadaqaCoins (New)

All praise belongs to Allah, and may His peace be upon the His messenger, his companions, and all of the believers.

The **SadaqaCoins Team** is pleased to present a gift and service for the Ummah. The believers now have the oppurtunity to advertise and support the latest projects of the Mujahedin.

A considerable advantage (over other platforms) is that it our service is anonymous and secure.

Please note that our website can only be accessed using the Tor Browser. However, do not let that put you off, as its Installation is simple, and then should ensure your online safety.

SadaqaCoins Darknet Website: (Only accessible via the Tor Browser!)
🔓 http://sadaqabmnor4ufnj.onion/

You can download the Tor Browser (Orbot+Orfox) on your smartphone from Google Play or F-Droid:

📲 Orbot:
https://play.google.com/store/apps/details?id=org.torproject.android

📲 Orfox:
https://play.google.com/store/apps/details?id=info.guardianproject.orfox

📲 Orbot & Orfox:
https://f-droid.org

For further questions, open a Telegram secret chat: @SadaqaCoins or contact us via email.

Follow us on Twitter https://twitter.com/SadaqaCoins, and share our Telegram Channel: @SadaqaCoinsNews

Email:
✉ 7363@tuta.io

👁 1144   edited 4:08 AM

On September 28, SadaqaCoins posted on Telegram an animated GIF showing how users can download and install F-droid. The second GIF posted from Orbot said: "You can enable any app to go through Tor using our built-in VPN [virtual private network]. This won't make you anonymous, but it will help get through firewalls."

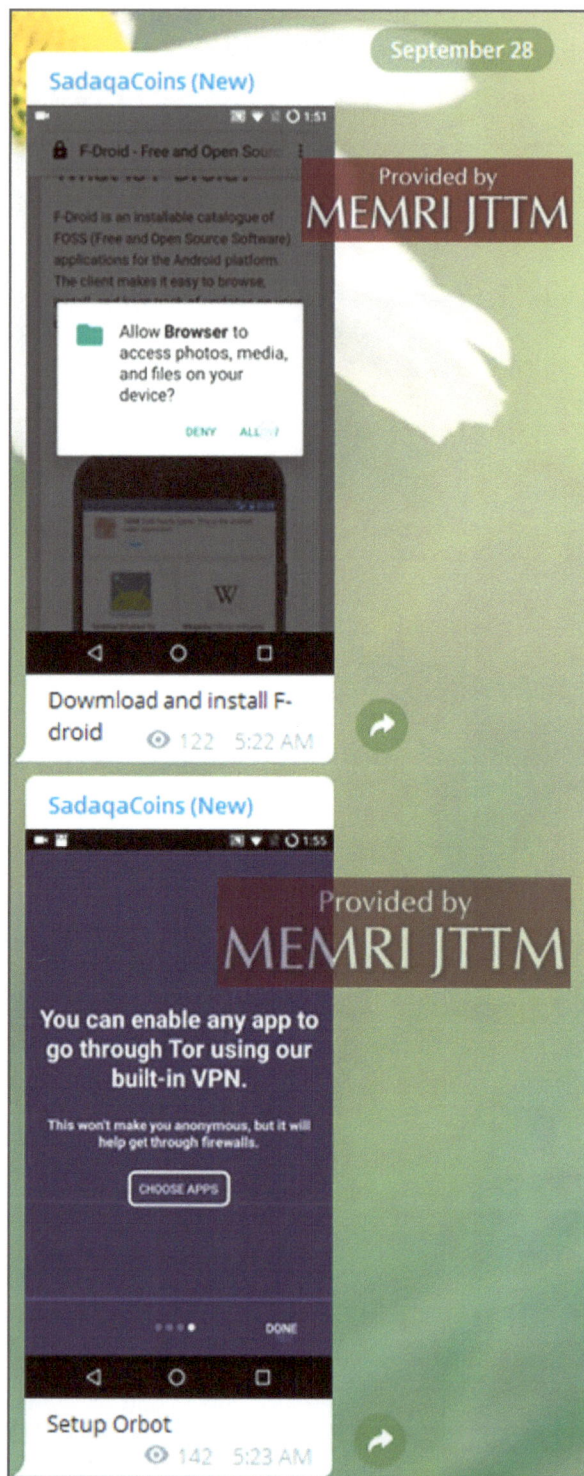

The Coming Storm: Terrorists Using Cryptocurrency

On September 29 and October 1, SadaqaCoins posted a link to its website and a link to a YouTube video from a YouTube Channel called HonkinNews. In one section of the 12-minute video, user Stephen E Arnold discusses SadaqaCoins.

September 29

SadaqaCoins (New)

موقع "صدقة كوينز" متوفر الآن باللغة العربية. يرجى نشر هذه الرسالة.

http://sadaqabmnor4ufnj.onion/ar/   👁 149   12:55 PM

October 1

SadaqaCoins (New)
#SadaqaCoins – activist Dark Web crowdfunding service
https://youtu.be/U-AsfRN8JEM?t=171   👁 75   12:13 PM

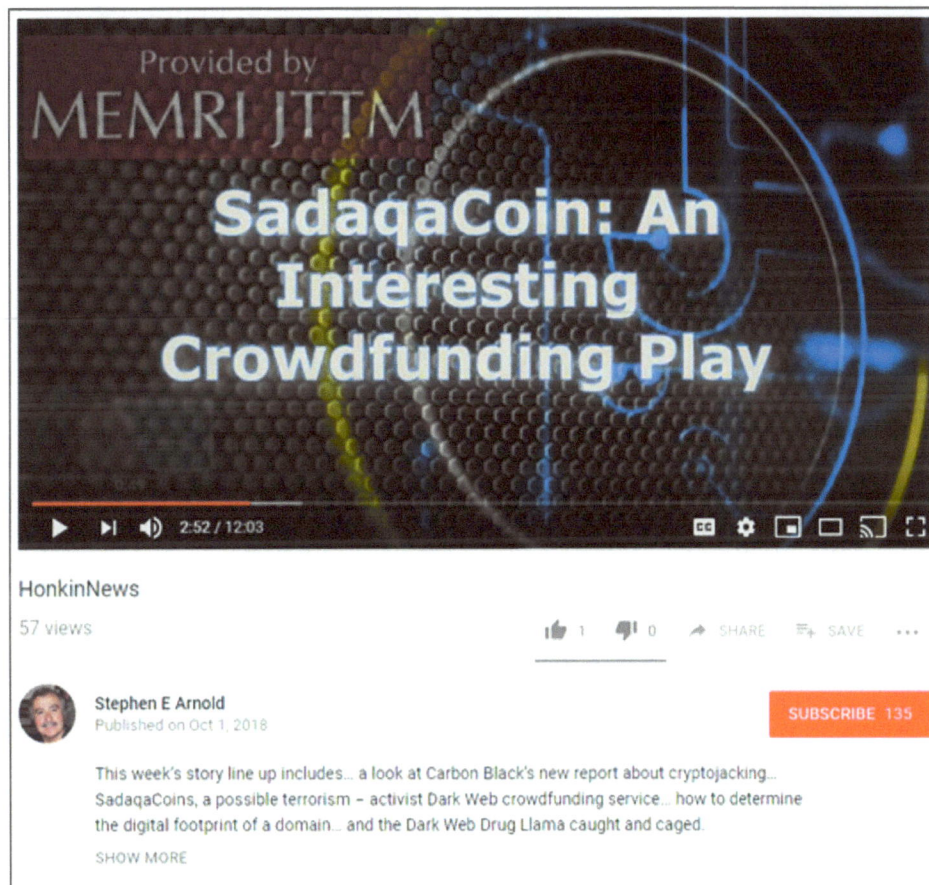

Provided by
MEMRI JTTM

SadaqaCoin: An Interesting Crowdfunding Play

▶ ▶| 🔊 2:52 / 12:03

HonkinNews

57 views   👍 1   👎 0   ↗ SHARE   ⊞+ SAVE   •••

Stephen E Arnold
Published on Oct 1, 2018   SUBSCRIBE 135

This week's story line up includes... a look at Carbon Black's new report about cryptojacking... SadaqaCoins, a possible terrorism – activist Dark Web crowdfunding service... how to determine the digital footprint of a domain... and the Dark Web Drug Llama caught and caged.

SHOW MORE

Jihadi Fundraising Group Al-Sadaqah Uses Cryptocurrencies: Activity November 2017-October 2018

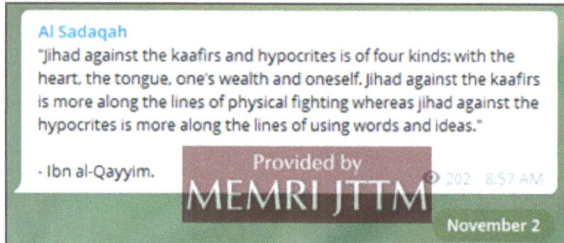

The Al-Sadaqah organization, which says it is a charity organization, raises funds for jihadi groups in Syria using the Bitcoin, Dash, Monero, and Verge cryptocurrencies, numerous social media platforms including Telegram, Facebook, Twitter, WhatsApp, and Instagram, the encrypted email service ProtonMail, the Deep Web-based email service SecMail, and its own website. The group solicits donations on social media, then later posts photos and videos of supplies, weapons, and construction materials purchased with the past donations to encourage future donations. No trace of the group prior to October 2017 has been found.

The group interacts with potential donors on social media, encouraging them to donate and teaching them how to use cryptocurrencies and other technologies. Sometimes this social media outreach is in private discussions on Telegram, while in other cases it is done openly on Twitter. The group looks for ways to improve its fundraising, such as encouraging users to send vouchers using Telegram's "Secret Chat" function, saying this method is a "new and completely anonymous" way to donate. The group also encourages donors to use the bitcoin ATMs located in many countries around the world to donate anonymously.

November 2017 – Al-Sadaqah Begins Fundraising For Jihad In Syria In English Using Cryptocurrencies

In November 2017, Al-Sadaqah ran an English-language fundraising campaign on social media platforms including Telegram and Facebook to raise money using cryptocurrencies from Western supporters to finance the jihad against the Assad regime in Syria.[38]

A poster from the November 2017 fundraising campaign included a quote by 13th-century Islamic scholar Ibn Taymiyyah: "Whoever is unable to take part in jihad physically but is able to take part in jihad by means of his wealth, is obliged to take part in jihad by means of his wealth. So those who are well off must spend for the sake of Allah."[39] The group uses this quote often in its campaigns.

The Coming Storm: Terrorists Using Cryptocurrency

The Al-Sadaqah Organization Online In November 2017

In November 2017 the group was active on a main Telegram channel[40] as well as on a Telegram chat group.[41] The group gave Telegram user @Alsadaqah_Admin as the contact point for questions.

In November 2017, the group shared 11 photos and one short video mentioning donating and fundraising. Images shared on the group's Telegram channel in late 2017 show what appear to be an entrenched position in a mountainous area. In discussions between supporters and members of the group on Telegram, the group claimed these images were taken in northeastern Syria, in the Latakia region.

Al Sadaqah
15 members

Provided by MEMRI JTTM

VIEW CHANNEL

Info

An independent charity organization that is benefiting and providing the Mujahidin in Syria with weapons, finical aid and other projects relating to the Jihad.

You can donate safely and securely with Bitcoin.

Fundraising Campaign

The group shared its Bitcoin wallet address for receiving funds.[42] The call for funding included a long message detailing the needs and objectives of the fundraising: "*NEW PROJECT* 1st stage. Upgrading one of the

Photos distributed by Al-Sadaqah.

frontline points on a strategic mountain top in [L]atakia, this point has been constantly being targeted by tank shells and sniper fire and in need of urgent repairs and upgrades before the winter season. Upgrades include: – Two brand new toilet facilities – Concrete floor and water drainage for the trench – Upgrading of 3 sleeping areas that would allow 7 brothers to sleep on rotation – Building of a completely new trench – Building of a completely new tunnel system – Upgrading kitchen and storage room [...] The new trench and tunnel system will allow the brothers to advance on their position to get a better look out on the enemy and prepare for attacks. [...] Total cost for the first stage of the project is only $750. Donate now anonymous[ly] with Bitcoin."

An Al-Sadaqah operative asks followers on Telegram to circulate the campaign poster on Facebook and other platforms.

Facebook accounts share a fundraising message and links to the Telegram group.

Contact With Potential Donors, Including Minors

In the group's Telegram channel and in other pro-Al-Qaeda chatrooms,[43] Al-Sadaqah operatives engaged in discussions with potential donors and offered advice, guidance, and incentives to donate to their cause. In two cases, an Al-Sadaqah operative engaged in discussions with minors claiming to be from Europe. In the first case, a 15-year-old from London[44] said that had tried in the past to join the fight in Syria and wanted to donate money to the mujahideen. In a second case, a 17-year-old from Germany[45] claimed that he had donated €500 after a discussion with a member of Al-Sadaqah.

Musa Muhammad stated that he is 15 and "was recently arrested and charged under section 5 [of the] terrorism act of 2006 in UK." He said that his father is a member of the Salafi group Hizb ut-Tahrir and that the only thing he wanted was "wanted martyrdom for the sake of Allah." He also expressed his desire to immigrate and join the jihad, and said that he had already attempted this and failed. The Al-Sadaqah members suggested to him that he should "stay in school [...] learn some Islamic knowledge. Then come to the next jihad Allah willing."

The Coming Storm: Terrorists Using Cryptocurrency

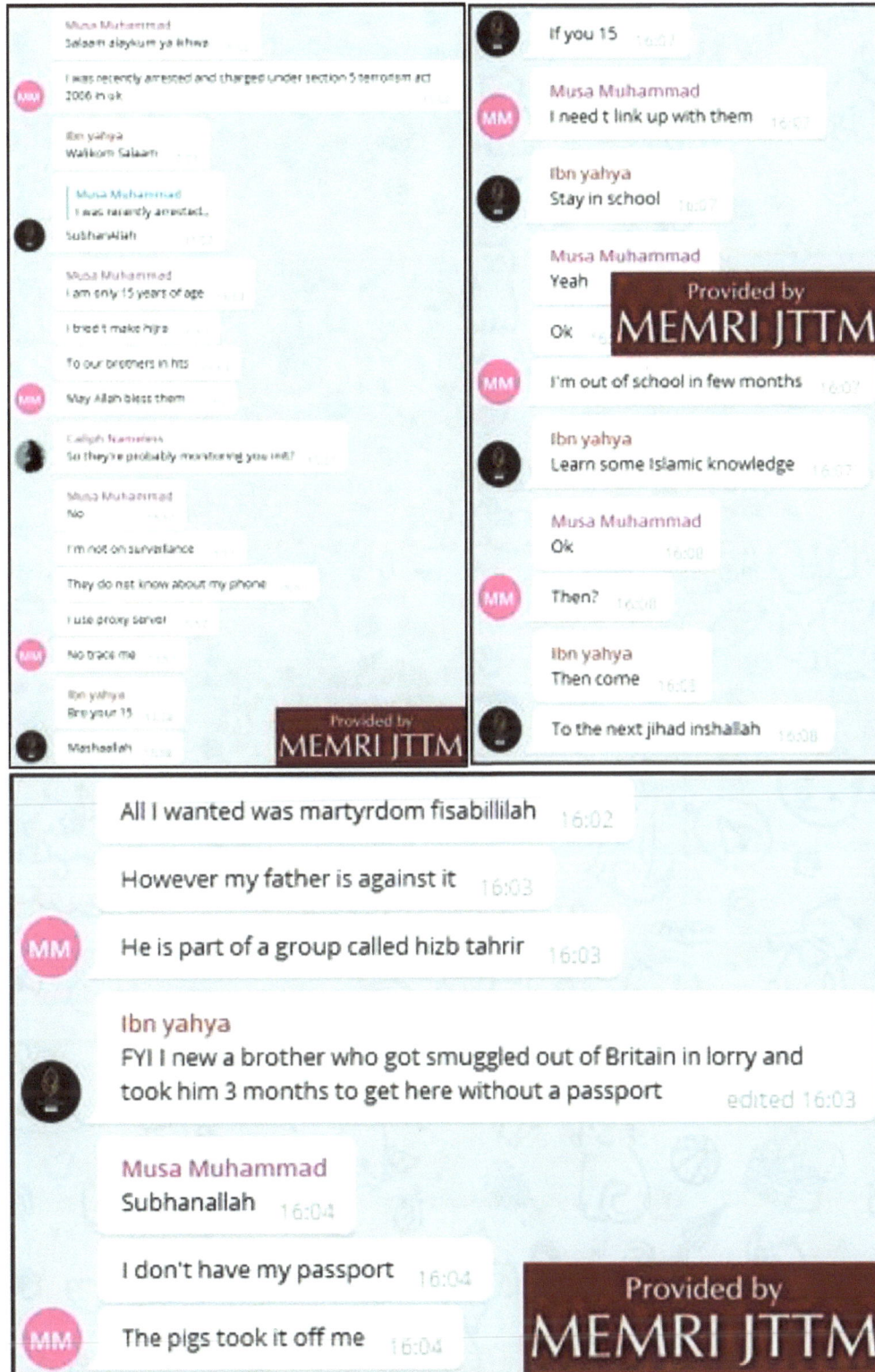

Musa Muhammad
Salaam aleykum ya khwa

I was recently arrested and charged under section 5 terrorism act 2006 in uk

Ibn yahya
Walikom Salaam

Musa Muhammad
I was recently arrested...
SubhanAllah

Musa Muhammad
I am only 15 years of age

I tried t make hijra

To our brothers in hts

May Allah bless them

Caliph Nameless
So they're probably monitoring you init?

Musa Muhammad
No

I'm not on surveillance

They do not know about my phone

I use proxy server

No trace me

Ibn yahya
Bro your 15

Mashaallah

If you 15

Musa Muhammad
I need t link up with them

Ibn yahya
Stay in school

Musa Muhammad
Yeah

Ok

I'm out of school in few months

Ibn yahya
Learn some Islamic knowledge

Musa Muhammad
Ok

Then?

Ibn yahya
Then come

To the next jihad inshallah

All I wanted was martyrdom fisabillilah 16:02

However my father is against it 16:03

He is part of a group called hizb tahrir 16:03

Ibn yahya
FYI I new a brother who got smuggled out of Britain in lorry and took him 3 months to get here without a passport   edited 16:03

Musa Muhammad
Subhanallah 16:04

I don't have my passport 16:04

The pigs took it off me 16:04

Al Sadaqah Admins instructs potential donors on how to buy Bitcoin.

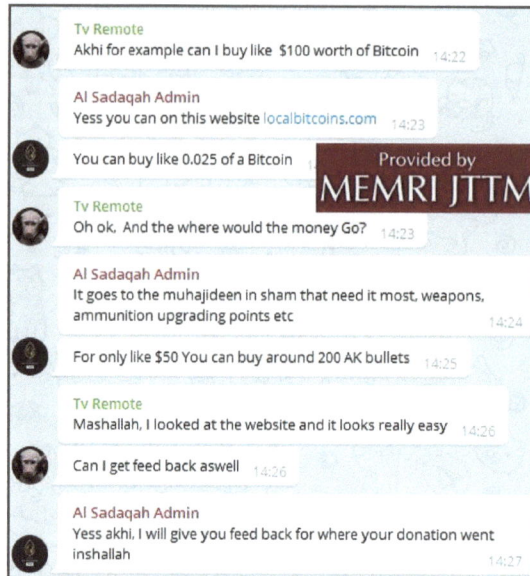

The Al-Sadaqah member encouraged and instructed a Telegram user who said he was 17 to "just put money in an envelope and send it."

The Coming Storm: Terrorists Using Cryptocurrency

December 2017 – Al-Sadaqah Solicits Funds On Telegram, Instagram In Bitcoin To Reinforce Military Facilities

On December 4, 2017, Al-Sadaqah circulated a graphic soliciting funds for the mujahideen in Syria. Proceeds from the fundraising campaign, titled "Project Re-Enforce" went toward reinforcing "defensive position with new trenches," and "guard posts with new tunnel system."[46] The graphic called upon contributors to donate $750, and a bitcoin address is provided on the graphic for donors. It appears that the fundraisers achieved their goal for this project and raised $875.66. Al-Sadaqah's Telegram byline at that time read: "Charity organization for helping Mujahideen in Syria. Donate securely through Bitcoin." The graphic was circulated on the Telegram page of a Pakistani HTS militant and recruiter.[47] The current Al-Sadaqah Telegram channel was created on December 4. It appears that the same fundraisers had also solicited donations during the previous month on a Telegram channel with the same name that was later shut down.[48]

The graphic read: "Winter is fast approaching and the mujahideen in Syria are holding on, steadfast with only one goal in their sights, eternal paradise. Along with the military might the *kuffar* [infidels] pose against them, the brothers are now facing a more difficult challenge, the harsh conditions of winter. Urgent re-enforcement of the Ummah's guarding posts are needed, who will respond?

"Re-enforce defensive position with new trenches.
"Re-enforce guard posts with new tunnel system.
"Re-enforce trench flooring to concrete.
"Re-enforce storage facilities and kitchen facilities.
"Re-enforce 3 vital resting areas, accommodates 7 brothers each.
"Re-enforce Toilet and bathroom facilities.
"Only $750."

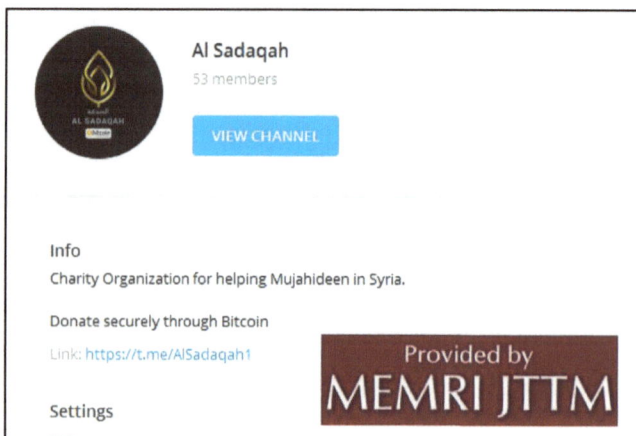

Al-Sadaqah Telegram channel.

The bitcoin address provided in the graphic is the same as the one listed by previous Al-Sadaqah Telegram channel in November. As of December 2017 there had been four transactions totaling $875.66 received. The group also provided a QR code.

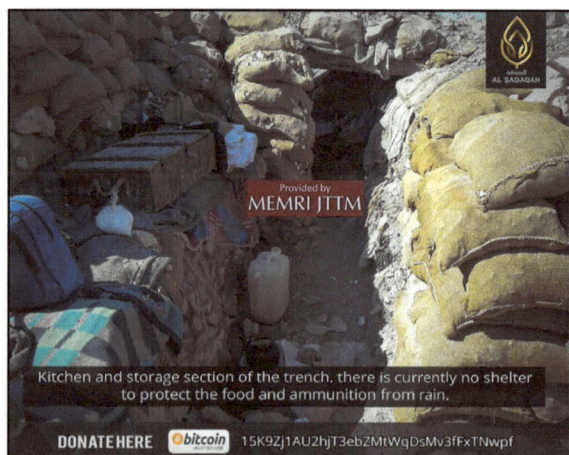

Another post showed the kitchen and storage section of a trench. The text read: "There is currently no shelter to protect the food and ammunition from rain."

In one post, Al-Sadaqah listed the Bitcoin address and noted that interested individuals can contact the Telegram channel's administrator if they wish to seek out other payment options.

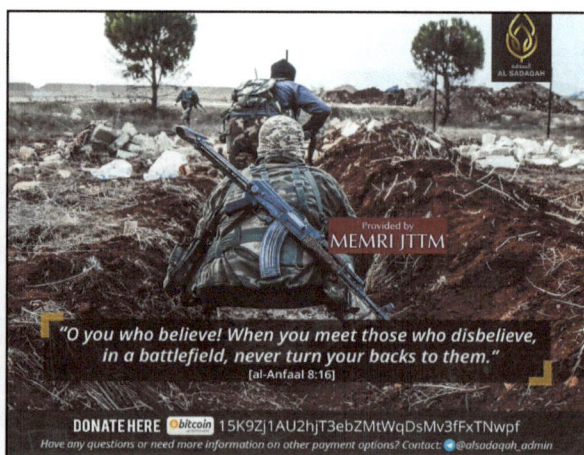

The Coming Storm: Terrorists Using Cryptocurrency

Al-Sadaqah also opened an Instagram account featuring the group's graphics.

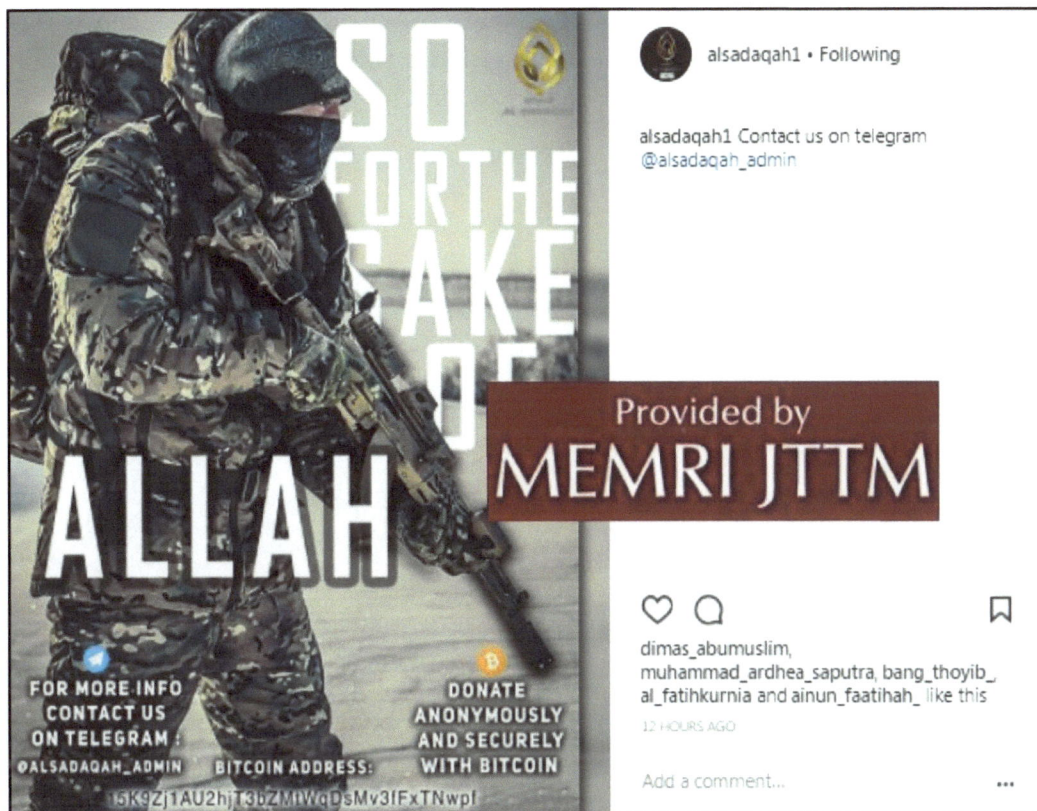

December 2017 – Al-Sadaqah Reports Successful Social Media Fundraising Of Bitcoin To Improve Conditions In The Trenches

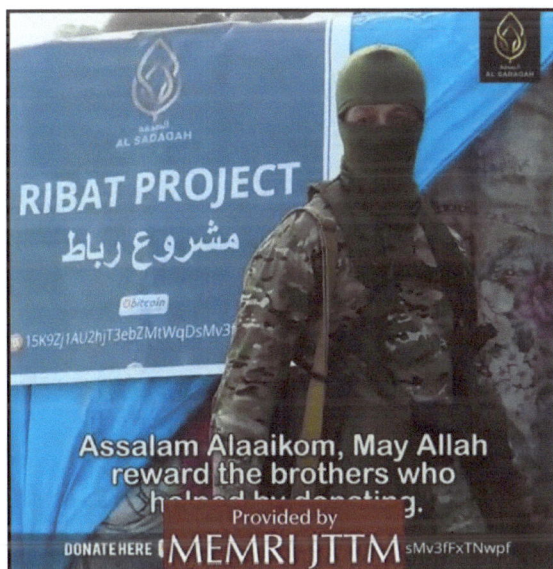

On December 16, 2017, the Al-Sadaqah organization posted a one-minute clip on Telegram showing a foreign fighter in a trench.[49] The fighter takes the viewer on a tour of what he describes as a frontline position where conditions were improved thanks to money raised online. The fighter addressing the camera speaks in Arabic in an American-sounding accent. The clip is part of the Al-Sadaqah organization "Ribat Project," which focuses on improving living conditions for a group of fighters in Syria. The organization is still active on Facebook,[50] Twitter,[51] Instagram,[52] and Telegram.[53]

The following is a transcription of the English subtitles provided in the clip:

"May Allah reward the brothers who helped by donating. This is the place where the brothers stay on *ribat* [frontline]. If you would come with me [the camera follows the fighter into a covered trench]. Alhamdulillah, this place where the brothers eat and sleep is in much better condition than it was before. All of this is thanks to Al-Sadaqah and your donations. Inshallah if you are able to continue to donate it would be very, very helpful."

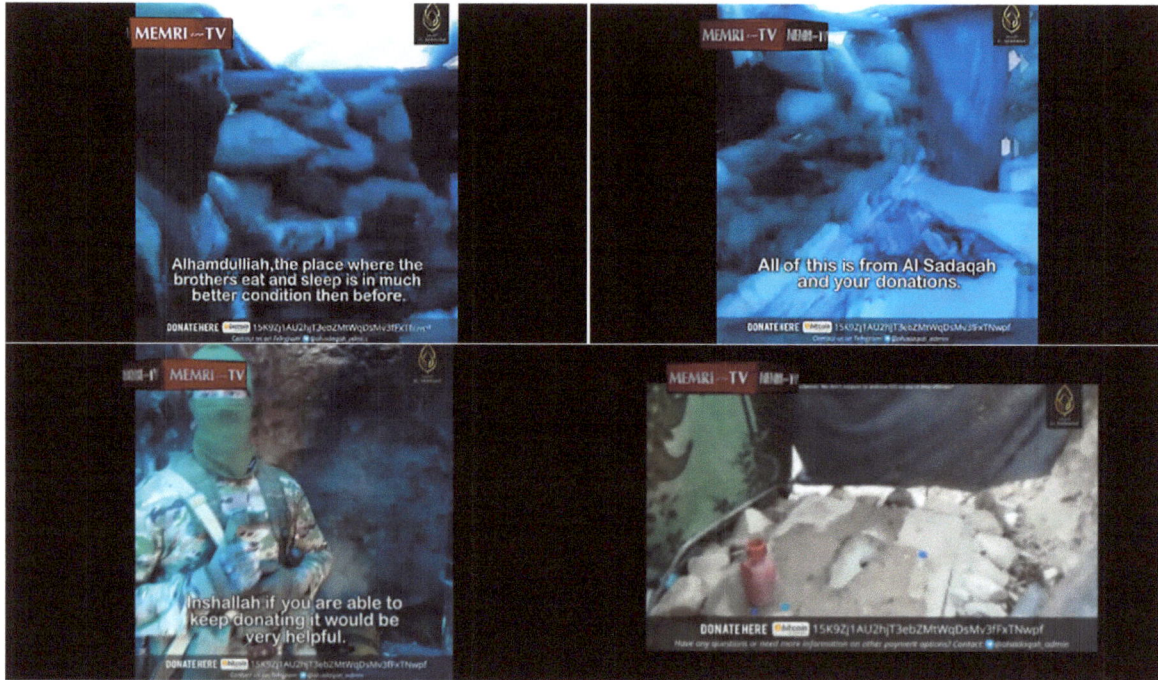

The video was accompanied by a post stating: "Allah be praised, your donations have been sent and used on the frontlines here in #Syria. May Allah reward you. Contact @alsadaqah_admin for donations."

Post on Telegram calling for donation through a bitcoin account.   *Al-Sadaqah profile on Twitter.*

The Coming Storm: Terrorists Using Cryptocurrency

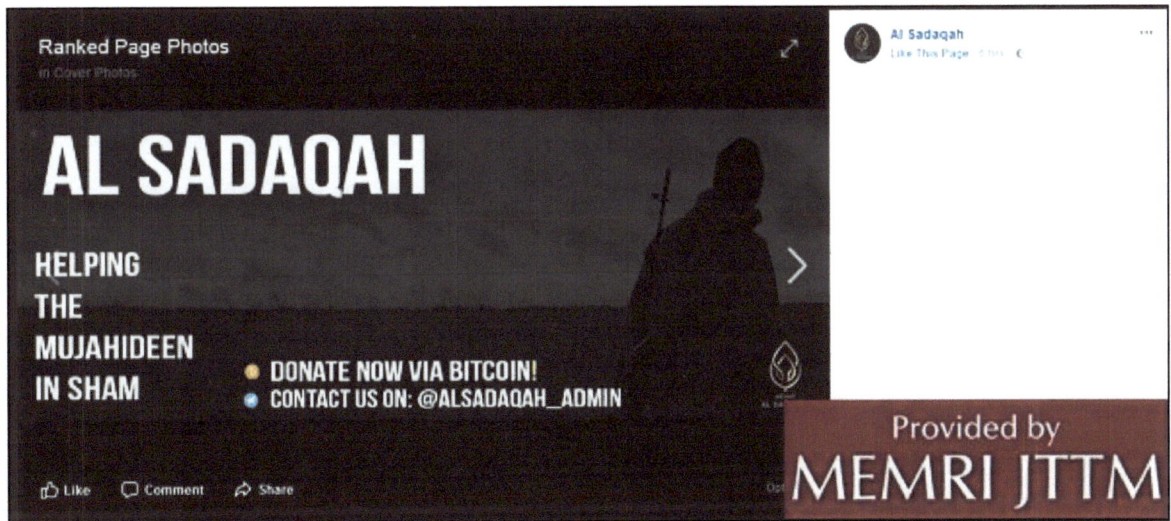

Facebook page of the group.

Instagram profile page.

December 2017 – Pakistani Engineer On Twitter With Al-Sadaqah: Give 800M Accurate Semi-Auto .338 Sniper Rifles With Thermal Sights, All Weather Uniform

On December 19, 2017, a Pakistani network engineer named Abdul Haseeb communicated on Twitter with the account of Al-Sadaqah.[54] According to his Twitter profile, Haseeb lives in Islamabad and works for Nayatel, a leading Pakistani Telecom company. At that time

Al-Sadaqah was promoting its campaign to raise funds via Bitcoin and Western Union on Telegram, Twitter, Instagram, and Facebook.

Provided by MEMRI JTTM

Abdul Haseeb

@Abdul_Haseeb92

Graduate Electronics Engineer, currently serving as Network Engineer at NAYAtel QA department.

⊙ Islamabad, Pakistan

📅 Joined October 2017

Al Sadaqah @AlSadaqah1 · Dec 19
Name some projects you would like to see done for the mujihideen and that you would be willing to sponser

#Syria #islam #assadbesiegesghouta

Abdul Haseeb @Abdul_Haseeb92 · Dec 19
Training to Malhama Tactical level for urban and gorilla warfare
800m accurate semi-automatic .338 sniper rifles with thermal sights
Standardized all weather uniform like uf-pro striker solution ufpro.com

Al Sadaqah
@AlSadaqah1

Provided by MEMRI JTTM

Replying to @Abdul_Haseeb92

Yess very good idea, biggest issue is training requires alot of discipline which unfournitly alot of the Syrian population don't have, getting all the equipment is a must and good idea. We will defs consider it InshaAllah

10:57 PM - 19 Dec 2017

1 Like

On February 9, Haseeb posted a photo of himself from his office.

On December 19, Al-Sadaqah tweeted: "Name some projects you would like to see done for the mujihideen and that you would be willing to sponser [sic]." Haseeb replied: "Training to Malhama Tactical level for urban and gorilla [sic] warfare 800m accurate semi-automatic .338 sniper rifles with thermal sights. Standardized all weather uniform like uf-pro striker solution ufpro.com." Al-Sadaqah responded: "Yes very good idea, biggest issue is training requires a lot of discipline which unfortunately a lot of Syrian population don't have, getting all the equipment is a must and good idea. We will defs consider it InshaAllah."

January 2018 – Al-Sadaqah Touts 'New And Completely Anonymous' Way To Donate In Bitcoin, Instructs Donors To Send Vouchers Via Telegram's Secret Chat

On January 16, 2018, the Al-Sadaqah organization posted an announcement on its Telegram channel about a new and anonymous method that people can use to send it money.[55] It wrote: "We have a new and completely anonymous way of donating through cash in most country's [sic]. Message @al-sadaqah11."[56]

The Coming Storm: Terrorists Using Cryptocurrency

A day later, Al-Sadaqah wrote: "Send us Bitcoin quick and easy, all you need to do is purchase a voucher and send us the code in secret chat on Telegram @alsadaqah11 ."

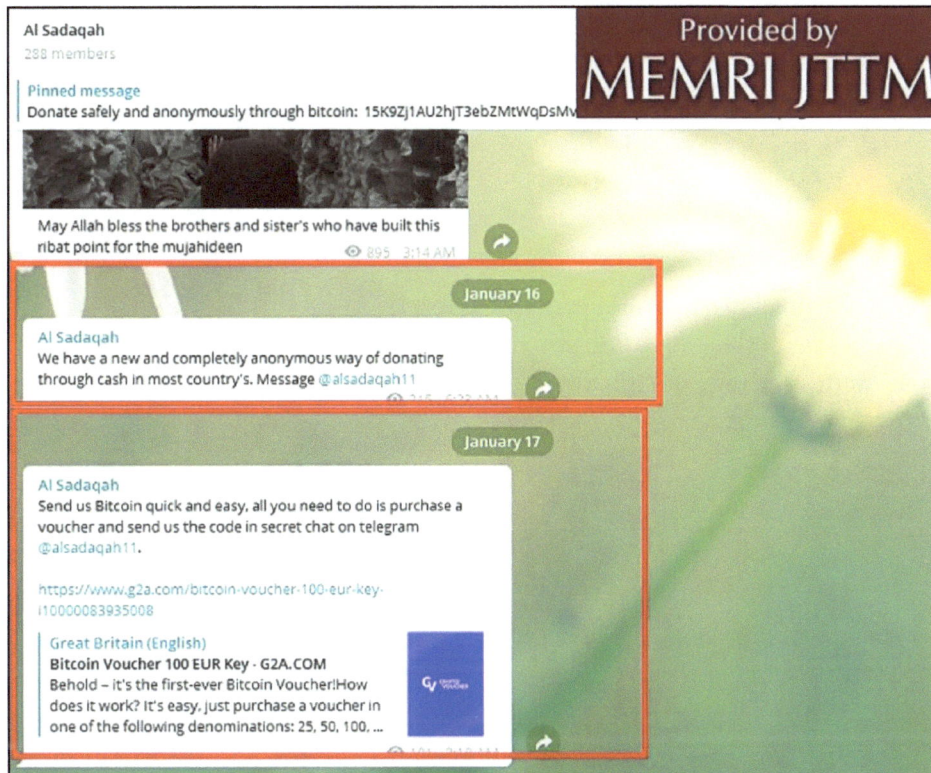

Al-Sadaqah Telegram announcements.

Al-Sadaqah directed users to the Crypto Voucher (cryptovoucher.io) website where people could purchase bitcoin vouchers, which are similar to gift cards. The website also allows people with a valid voucher number to redeem their card and transfer the funds into their respective bitcoin wallet. That voucher number is the same number that Al-Sadaqah asked people to send it to directly via Telegram's end-to-end encrypted secret chat.

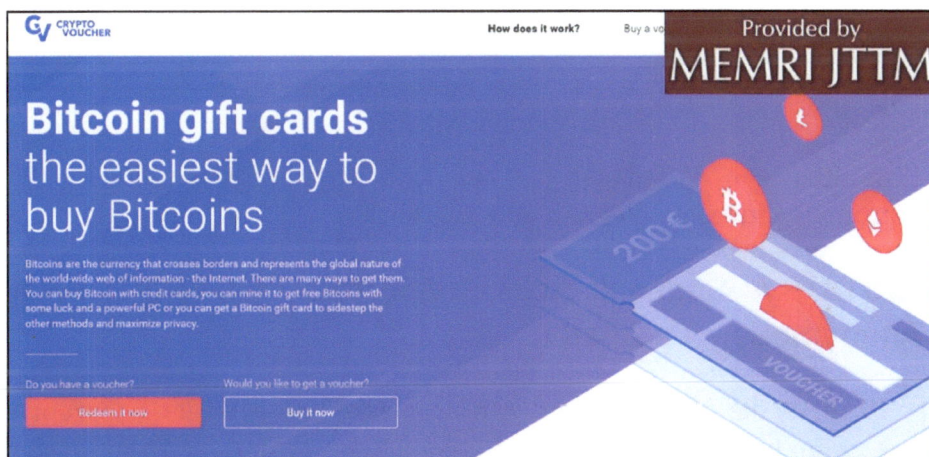

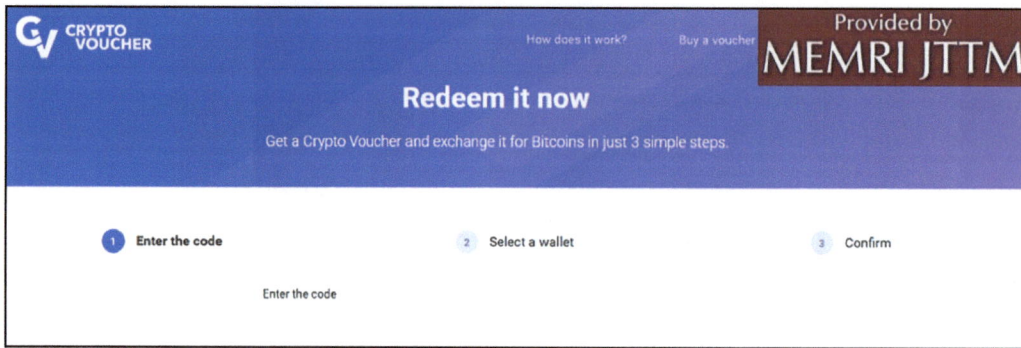

Vouchers can be purchased for amounts ranging from £25 to £1,000 British. The transactions themselves take place on another website, G2A.com, which provides people with six different payment methods during checkout.

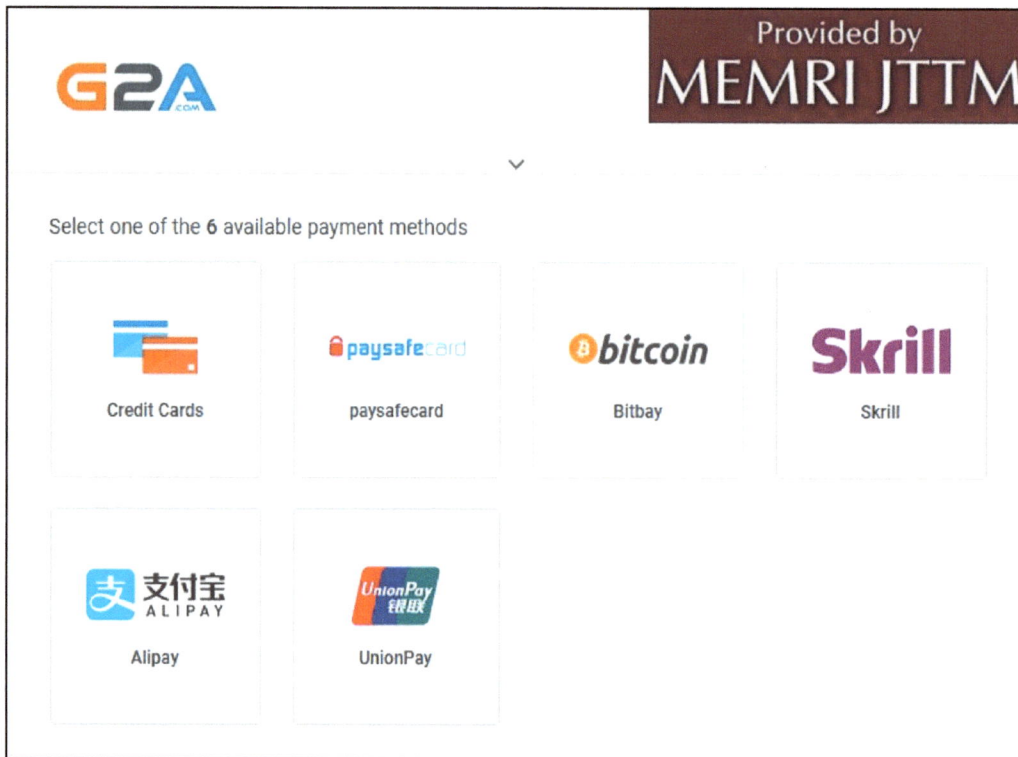

Available payment methods to purchasing bitcoin voucher.

For several months now, Al-Sadaqah has been soliciting funds via Bitcoin and Western Union. The group is active on Telegram and Twitter.[57]

January 2018 – Al-Sadaqah Asks Supporters To Use Bitcoin ATMs To Donate Anonymously To Jihad

On January 19, 2018, the Al-Sadaqah organization asked users on its Telegram channel to donate to the group using the bitcoin automated teller machines (ATMs) located in many countries around the world.[58]

The Coming Storm: Terrorists Using Cryptocurrency

The announcement read: "If anyone has a Bitcoin ATM in your area or country, then you can send money to the mujhideen [sic] 100% anonymously with cash. It is really that simple."[59]

Al-Sadaqah gave a link to the website Coinatmradar.com for a list of all available cryptocurrency ATMs worldwide and asked people to contact the website's administrator directly on Telegram for help.

Al-Sadaqah bitcoin ATMs announcement.

Map showing locations of bitcoin ATMs and other related services worldwide (source: coinatmradar.com.)

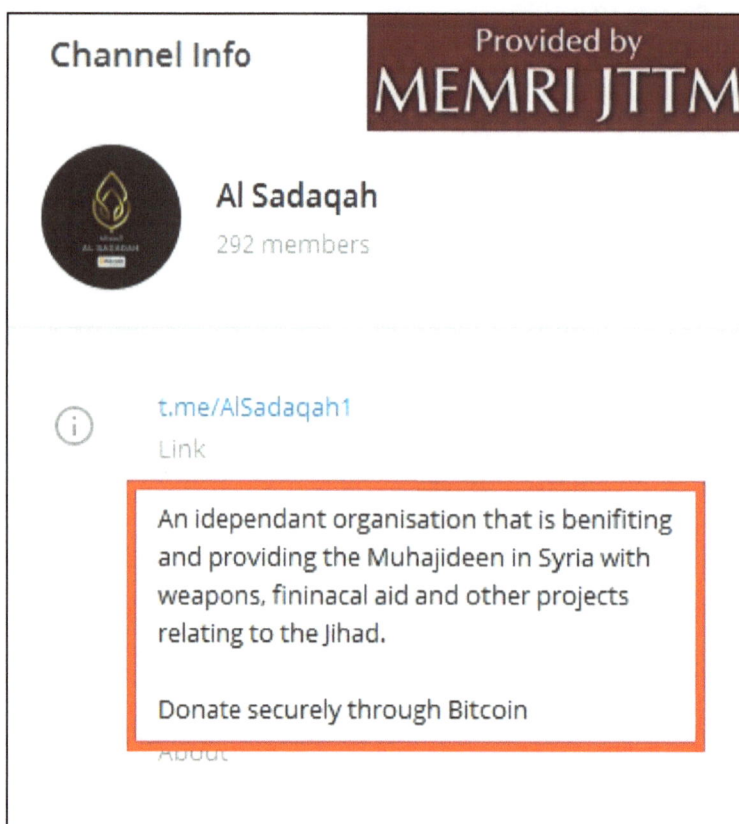

Al-Sadaqah's Telegram channel noting the group's pro-jihad work.

February 2018 – Al-Sadaqah Organization Launches Website, Promotes Use Of Bitcoin, Other Cryptocurrencies To 'Sponsor' Mujahideen, Purchase Weapons In Syria

In February 2018, Al-Sadaqah organization opened a website. The website offered donation options using various cryptocurrencies, which were used to "sponsor" a fighter in Syria and to procure weapons and other battle equipment, among other things.[60] The Al-Sadaqah website complements the group's fundraising efforts on Telegram and Twitter.[61]

Al-Sadaqah had at that time opened a new Telegram channel after its previous one was deleted. On February 9, Al-Sadaqah announced on its Twitter account that its Telegram channel had been deleted.[62] A day later, a new Telegram channel was opened.[63]

Al-Sadaqah on Twitter noting its Telegram channel had been deleted.

The Coming Storm: Terrorists Using Cryptocurrency

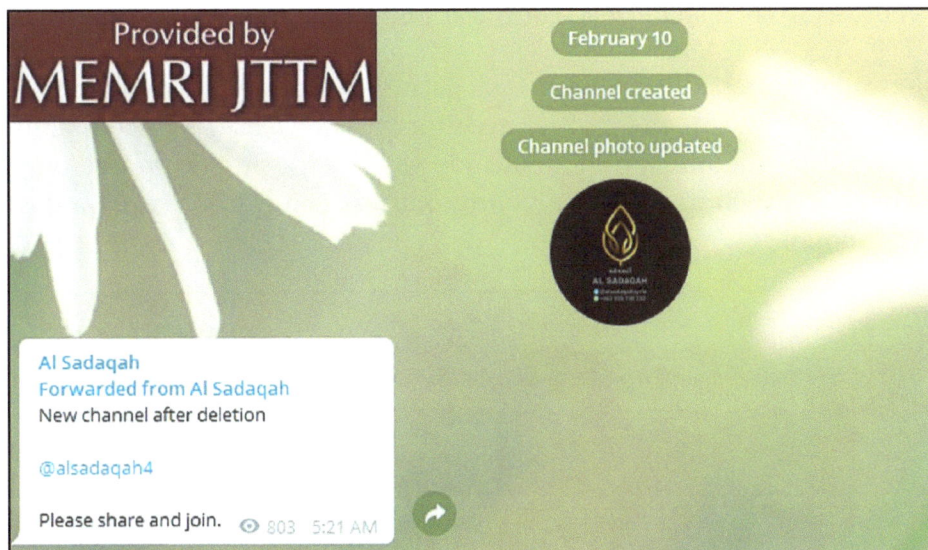

Al-Sadaqah promoting its new Telegram channel.

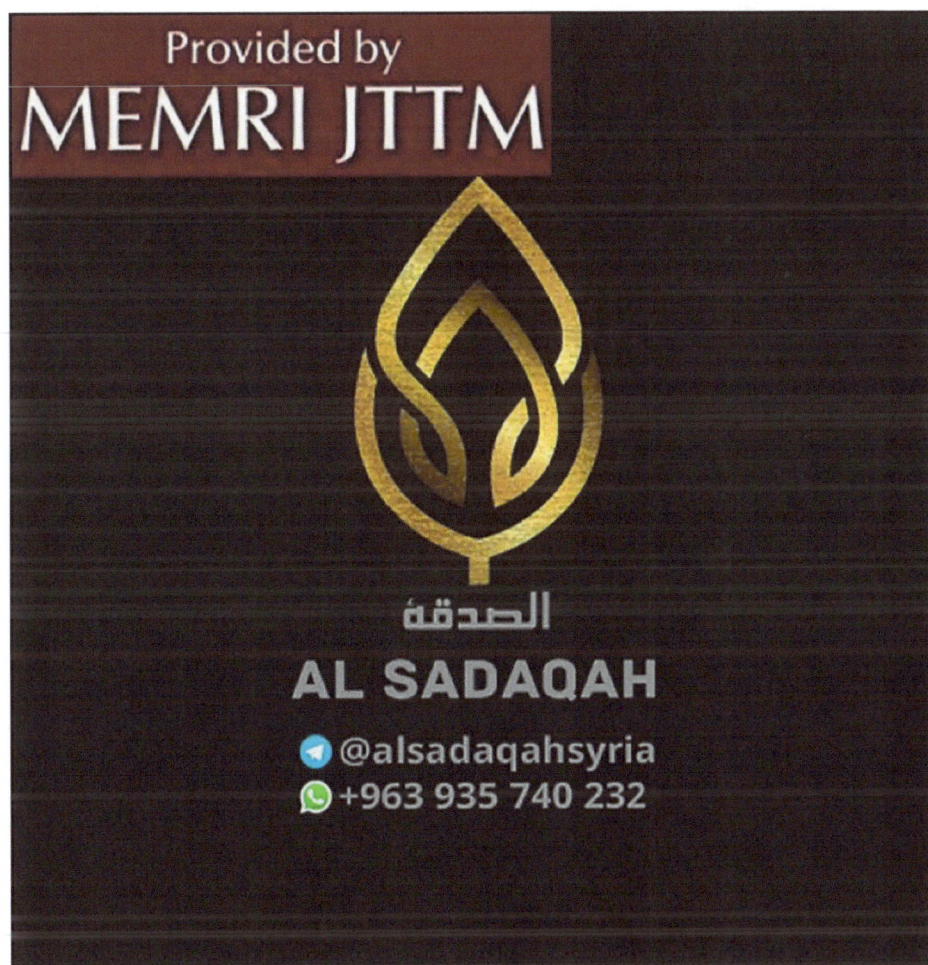

Al-Sadaqah's admin Telegram handle (@alsadaqahsyria) and WhatsApp number (+963935740 232).

The new website (al-sadaqah.mozello.com) was built using the website builder Mozello.[64]

The bitcoin wallet address[65] listed on Al-Sadaqah's website was different from the one listed on the group's Telegram and Twitter accounts.[66] The website's bitcoin address shows no transactions.[67] Furthermore, the Al-Sadaqah website was not promoted on the group's Telegram channel nor on its Twitter account, but was confirmed via private channels by Al-Sadaqah's admin.[68]

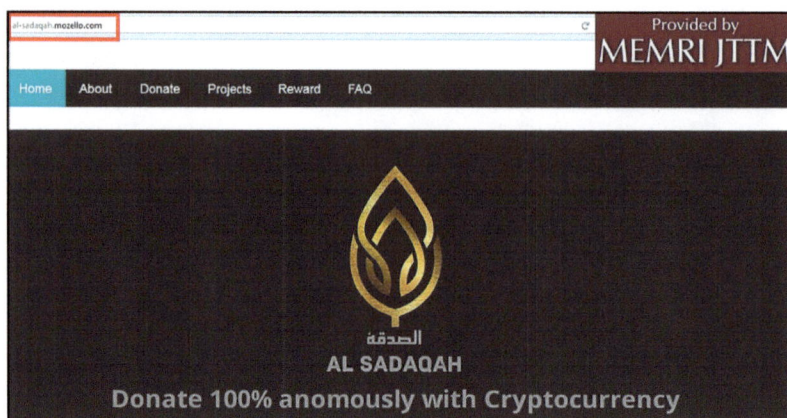

Al-Sadaqah website.

On its About page, Al-Sadaqah wrote: "Our aim is to assist all mujahidin that are fighting and defending the front lines in Syria. Since 2011, the brave Mujahidin have been battling the Syrian regime in order to stop the oppression and injustice happening to our Muslim brothers and sisters. Our Muslim woman and children have been imprisoned, shot, killed and raped by this barbaric regime and only a few chosen people of Ummah are defending their rights and our religion... The Jihad is Syria constantly needs your donations to survive and to continue the fight."[69]

The donate page listed several cryptocurrency options using which donations could be sent, including Bitcoin, Dash, Monero, and Verge. Only the bitcoin wallet address was listed; users need to contact Al-Sadaqah for information on all other cryptocurrencies.

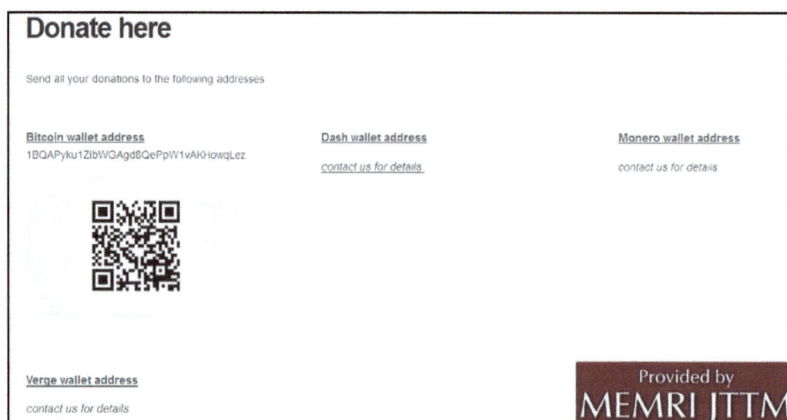

Al-Sadaqah website donation page.

The Coming Storm: Terrorists Using Cryptocurrency

To further obscure the source of donations, the website suggested that bitcoin users use at least two bitcoin addresses, moving the funds from one to the next before sending it to the group. "This breaks the chain and there is no way they [i.e., law enforcement agencies] can prove it was you," the website says.[70]

The website noted that the safest way to send money "without getting caught" was to use the Dash or Monero cryptocurrencies: "Dash has a special anonymous feature that allows it to be sent undetected without being seen by your government."[71]

The website listed the various "projects" that the allocated funds sent to the group are used for, including "sponsor a mujahid," which could be done for as little as $50 a month. It said: "Syria's ongoing conflict, now in its 7th year, has put a massive strain on the mujahid and his family. Most of the mujahedin that are defending the front line or are in the battles defending the Ummah receive little to no help financially, this gives the brothers no chance but to leave the front line to find work to support himself and his family. You can sponsor your own mujahid for as little as $50 a month. Your donation will provide him a wage that will in able him to stay on the front line knowing that he is able to provide for himself and his family."[72]

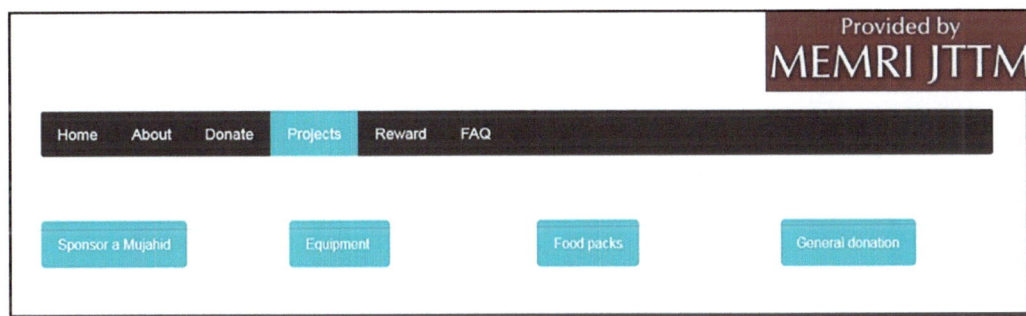

Al-Sadaqah's "projects."

Cost*	Mujahidin	Months
$50	1	1
$100	2	1
$300	1	6
$600	1	12

*US dollars

Suggested monthly "Sponsoring" cost for a mujahid.[73]

Other projects listed on the website were for buying weapons, various kinds of combat gear, and equipment for the mujahideen. The website said: "The most important thing for a mujahid when he is fighting his enemies is to have the right camouflage and equipment. Unfortunately, majority of the brothers who are defending the front line are under equipped with the most basic essentials. Some are unable to afford simple military camouflage which is vital for any soldier. Providing the most basic uniform and equipment will prepare the mujahid and give him confidence while stationed on ribat [guarding and fortifying territory] and also in battle. With your donation you can cover the following for a mujahid."[74]

Equipment

- Full military uniform
- Army boots
- Army gloves
- Army mask
- Army belt

Equipment

- Full military uniform
- Army gloves
- Ak47 and magazines
- Army knife
- Army vest
- Army jacket
- Army boots
- Army back pack
- Arm
- Arm

Provided by MEMRI JTTM

Some of the equipment that are needed for the mujahideen.[75]

Al-Sadaqah shared its Monero wallet address on social media so those interested could anonymously "donate now to support the brothers on the front lines."

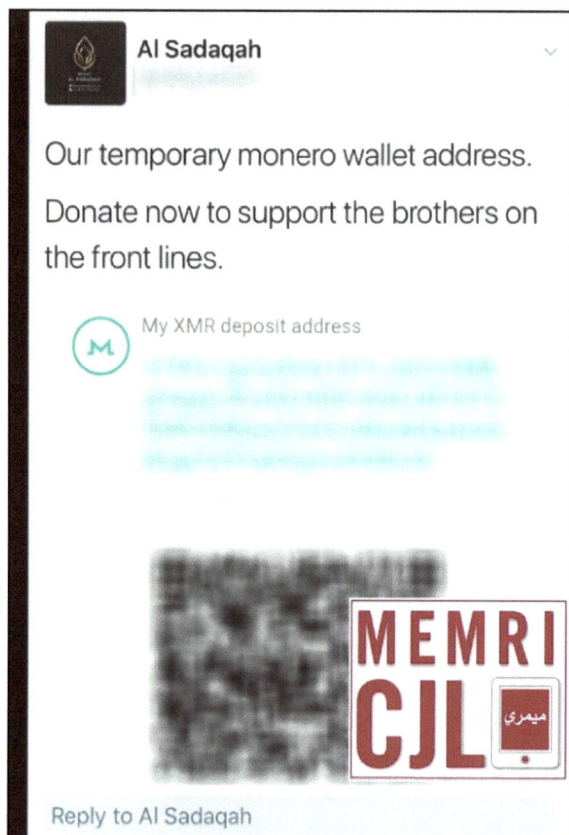

Our temporary monero wallet address.

Donate now to support the brothers on the front lines.

My XMR deposit address

May-June 2018 – A Review Of Al-Sadaqah Organization's Activity On Telegram: Group Updates Its WhatsApp Number, Uses Deep Web-Based Email Service SecMail

Though the Al-Sadaqah's Telegram channel has been shut down at least once, the group continued to operate on new channels in the summer of 2018.[76] The group can also be contacted via a WhatsApp number provided in May 2018 (see image below). In June 2018, Al-Sadaqah switched its contact email from the encrypted email service ProtonMail to the deep web-based SecMail.[77] According to

The Coming Storm: Terrorists Using Cryptocurrency

SecMail's website, it is an email service that is "fully established in the deep web," and which promises its users anonymity, saying: "You can be sure that nobody knows your identity."[78]

Al-Sadaqah's new WhatsApp number from May 2018 (left), and its previous number from February 2018 (right).

The following is a review of the content posted on the most recent Al-Sadaqah Telegram channel,[79] which the group opened on May 18, 2018. The review covers 16 photos, all of which urge people to donate to the group. The contact information shared on the channel includes: WhatsApp (+963 998 050 987), ProtonMail (alsadaqah@pm.me), and SecMail (alsadaqah@secmail.pro). The Bitcoin address 15K9Zj1AU2hjT3ebZMtWqDsMv3fFxTNwpf was also provided.

On May 18, 2018, the channel posted a photo showing two masked men wearing military fatigues and carrying assault rifles in a mountainous area. The two men in the picture hold up a sign that says "Ribat Project" and bears the emblem of the Al-Sadaqah Foundation and gives the group's Bitcoin address. The text on the photo reads: "Delivering food and supplies to the frontlines in Syria."

On May 25, 2018, the channel shared a poster saying Al-Sadaqah is providing a Ramadan *if-tar* "for mujahideen stationed on the frontline in Syria." A table estimates that $6,000 would benefit 100 fighters for 30 days; $1,800 would benefit 30 fighters for 30 days; $600 would benefit one fighter for 30 days; and $2 would benefit one fighter for one day. The poster gives the price of a pair of combat boots as $50 and the price of a military uniform as $50, and tells donors that the donation is a valid form of *zakat* ("alms-giving").

On June 1, 2018, the channel posted a photo showing six packaged military uniforms, each topped by a picture bearing Al-Sadaqah's emblem, a call for donations, and the Al-Sadaqah Bitcoin account number. The text on the photo read: "providing brand new military suits [uniforms] to mujahideen."

On June 1, 2018, the channel posted a photo showing seven mujahideen wearing military fatigues, their faces digitally blurred, each holding one of the packages shown in the previous photo. The text on the photo reads: "Providing brand new military suits to mujahideen."

The Coming Storm: Terrorists Using Cryptocurrency

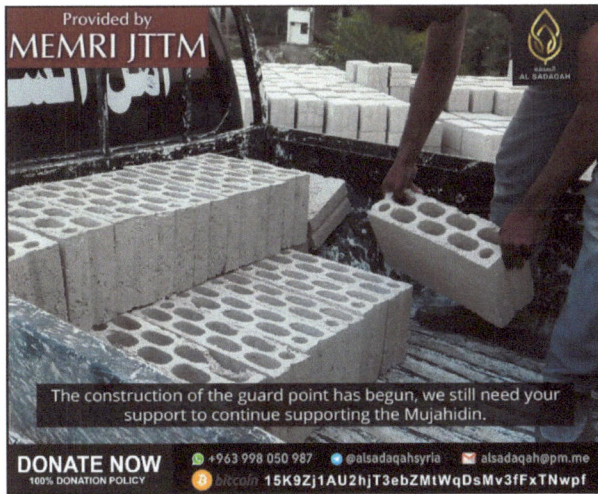

The construction of the guard point has begun, we still need your support to continue supporting the Mujahidin.

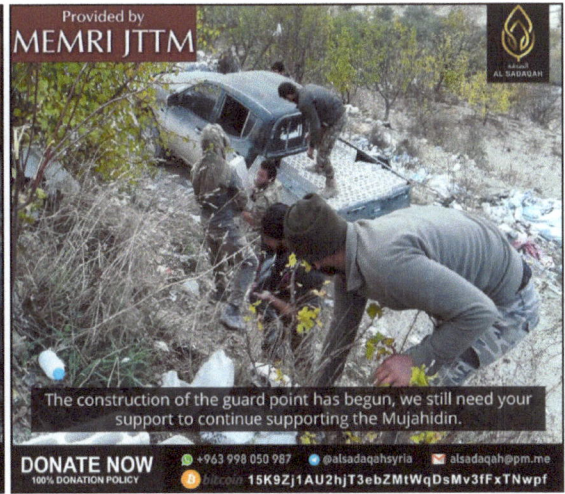

The construction of the guard point has begun, we still need your support to continue supporting the Mujahidin.

On June 9, 2018, the channel posted a photo of a man loading cinderblocks onto a pickup truck, with many cinder blocks in the background. The text on the photo reads: "The construction of the guard point has begun, we still need your support to continue supporting the Mujahidin."

On June 9, 2018, the channel posted a photo showing mujahideen in fatigues unloading cinder blocks from a pickup truck in a mountainous area. The text on the photo reads: "The construction of the guard point has begun, we still need your support to continue supporting the mujahidin."

On June 10, 2018, the channel posted a photo showing a mujahid, whose face was digitally blurred, carrying a cardboard box. The text on the photo read: "Supplying food rations to mujhaideen at the front lines."

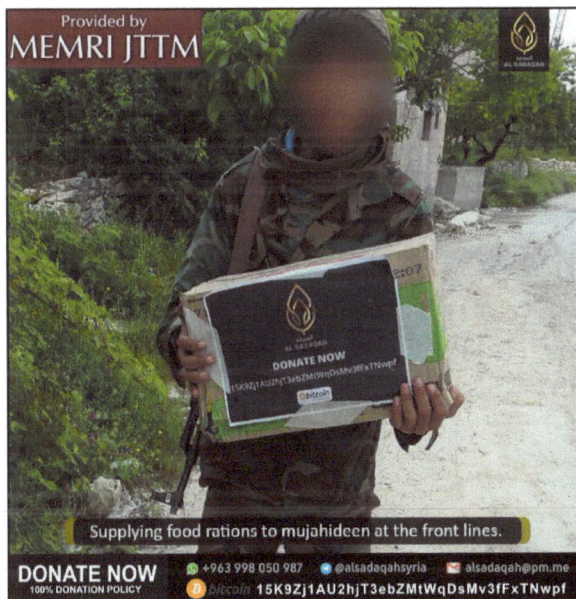

Supplying food rations to mujahideen at the front lines.

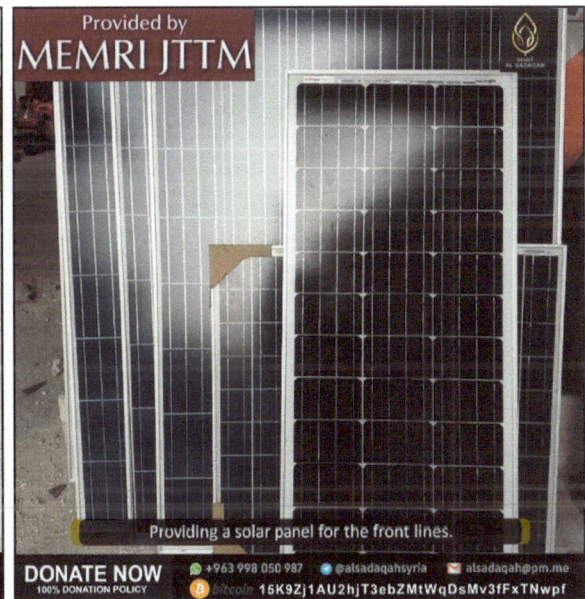

Providing a solar panel for the front lines.

Stalinsky ■ M E M R I

Providing a battery for the front lines.

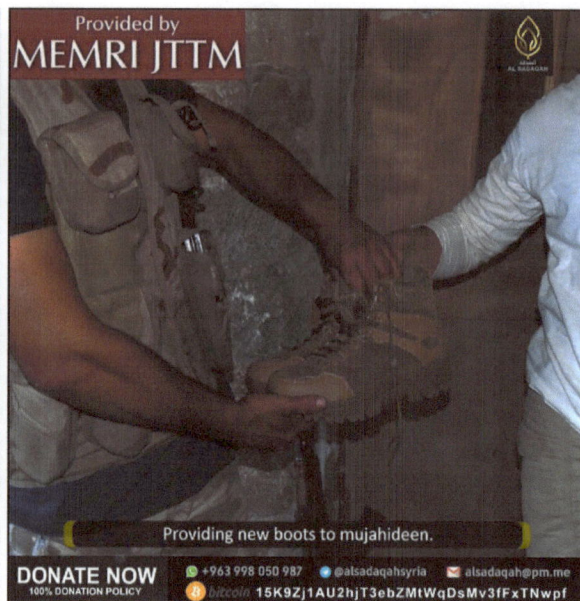
Providing new boots to mujahideen.

On June 12, 2018, the channel posted a photo of solar panels. The text on the photo read: "Providing a solar panel for the front lines."

On June 13, 2018, the channel posted a photo showing an assortment of batteries and boxed chargers. The text on the photo read: "Providing a battery for the front lines."

On June 14, 2018, the channel posted a photo showing a mujahid receiving a pair of combat boots. The text on the photo read: "Providing new boots to the mujahideen."

On June 18, 2018, the channel posted a photo showing two mujahideen, whose faces are digitally blurred, operating a mortar in a wooded area. The text on the photo read: "Financing and supporting the mujahideen in Syria."

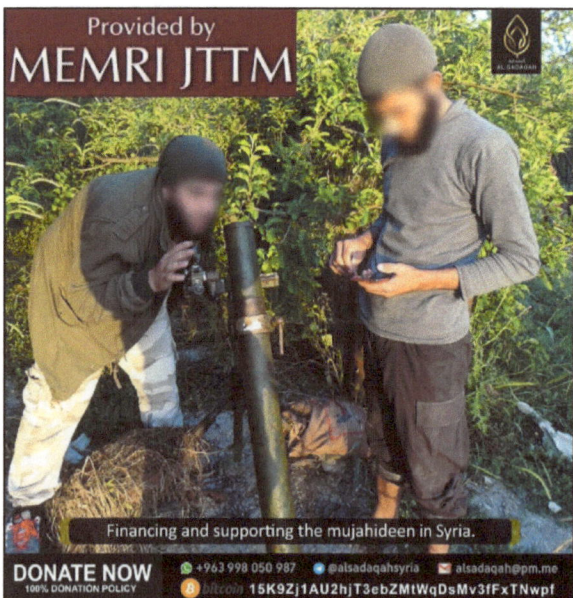
Financing and supporting the mujahideen in Syria.

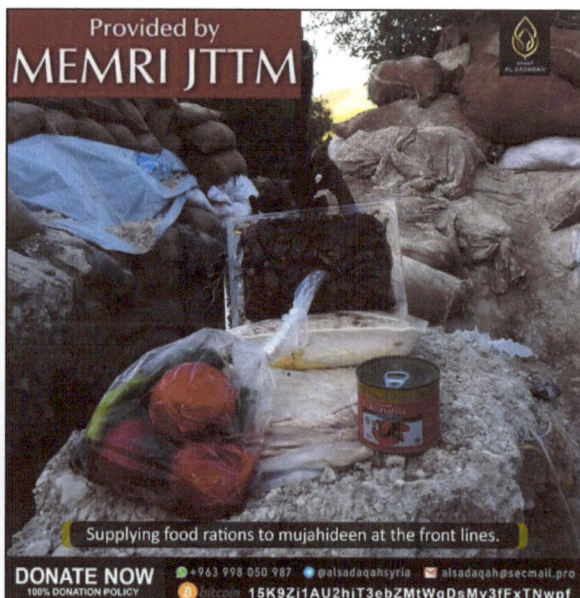
Supplying food rations to mujahideen at the front lines.

The Coming Storm: Terrorists Using Cryptocurrency

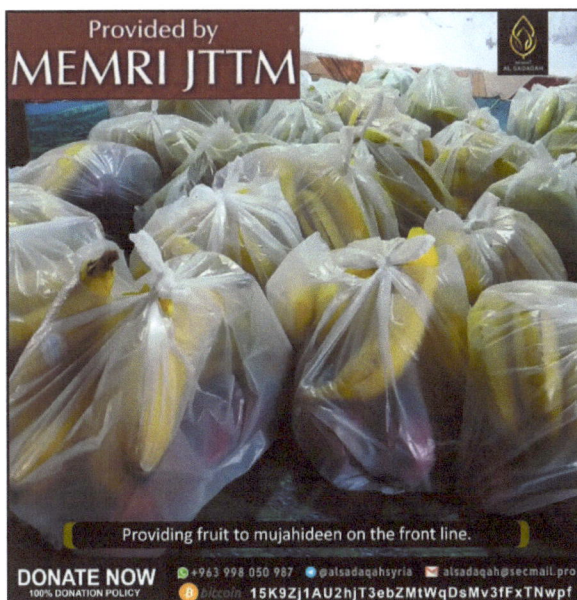
Providing fruit to mujahideen on the front line.

Providing food supplies to mujahideen.

On June 24, 2018, the channel posted a photo showing an enclosure surrounded by sand-bags and other fortifications. In the center is a box of dates leaning against a machine gun as well as other food. The text on the photo read: "Supplying food rations to mujahideen at the front lines."

On June 25, 2018, the channel posted a photo showing bags full of fruit. The text on the photo read: "Providing fruit to mujahideen on the front line."

On June 26, 2018, the channel posted a photo showing a muja-hid, whose face is digitally blurred, standing behind a pickup truck loaded with boxes, some of which bear the Al-Sadaqah emblem. The picture was taken in a wood-ed area. Another mujahid, whose face is digitally blurred, stands in the background. The text on the photo read: "Providing food sup-plies to mujahideen" along with a bitcoin wallet number for dona-tions.

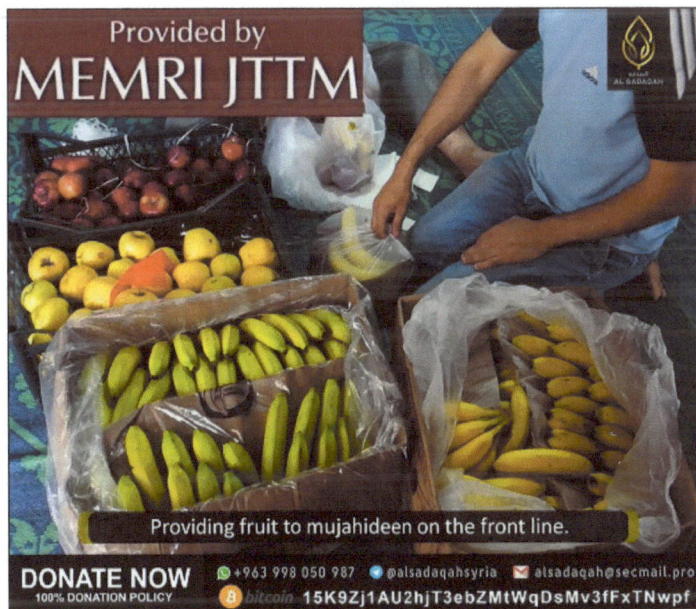
Providing fruit to mujahideen on the front line.

On June 30, 2018, the channel posted a photo of two men putting fruit into bags. The text on the photo read: "Providing fruit to mujahideen on the front line."

The Coming Storm: Terrorists Using Cryptocurrency

On August 11, 2018, a graphic designer based in Syria who is active on Telegram[80] published a poster in support of a campaign led by Al-Sadaqah.[81] The poster calls on Muslims to support jihadi fighters in Syria by donating funds that "will help us [Al-Sadaqah] equip frontline posts with specially designed mountain bikes to enable them to transport reinforcements and military supplies..." The poster includes contact details such as Al-Sadaqah's new WhatsApp number and its Bitcoin wallet number.

The release of the poster followed a post shared in English on the Al-Sadaqah Telegram channel[82] on July 2, 2018, that read: "There is a small group of Islamic mujahideen who are currently based in the mountains of Turkman in northern Syria. They have been doing ribat for over 4 months and are in desperate need of a mountain bike so they are able to travel back and forth to get supplies with ease. We can't stress the importance of having reliable transport back and forth from the front line. The main purpose of this bike would be to bring brothers back and forth to the front line, deliver food, water and ammunition supplies and for upcoming battles. The cost of a brand-new bike is only $950 and will be a massive benefit and ease of hardship for these brothers in ribat. The amount of [divine] reward you would receive is beyond comprehension inshallah. [...] Please don't hesitate to contact us if you would like to donate or have any questions."

The message and the poster provide the following contact details for potential donors:

Telegram: Telegram.me/alsadaqahsyria

WhatsApp number: +963 998 050 987

Email addresses: alsadaqah@secmail.pro and alsadaqah@pm.me

Bitcoin wallet address: 15K9Zj1AU2hjT3ebZMtWqDsMv3fFxTNwpf

DONATE NOW →
15K9Zj1AU2hjT3ebZMtWqDsMv3fFxTNwpf
alsadaqah@pm.me
+963 998 050 987
@alsadaqahsyria
Provided by
MEMRI JTTM

October 2018 – Al-Sadaqah Announces New Facebook Page, Gives WhatsApp Number, Bitcoin Address

In October 2018, Al-Sadaqah posted a message on its Telegram account announcing that it had a new Facebook page after the previous one was deleted.[83] Al-Sadaqah can also be contacted via WhatsApp and email. In June 2018, Al-Sadaqah switched its contact email from the encrypted email service ProtonMail to the deep web-based SecMail. According to SecMail's website, it is an email service that is 'fully established in the deep web,' and which promises its users anonymity."[84] Al-Sadaqah had posted most of the images in this report on its past accounts and here has reposted them on its new Facebook page.

On October 29, 2018, Al-Sadaqah wrote on its Telegram channel: "Please link and share our new facebook page after previous one was deleted." The post included a link to the new Facebook page.

As of this writing, the new Facebook page had 18 "likes" and 19 followers.

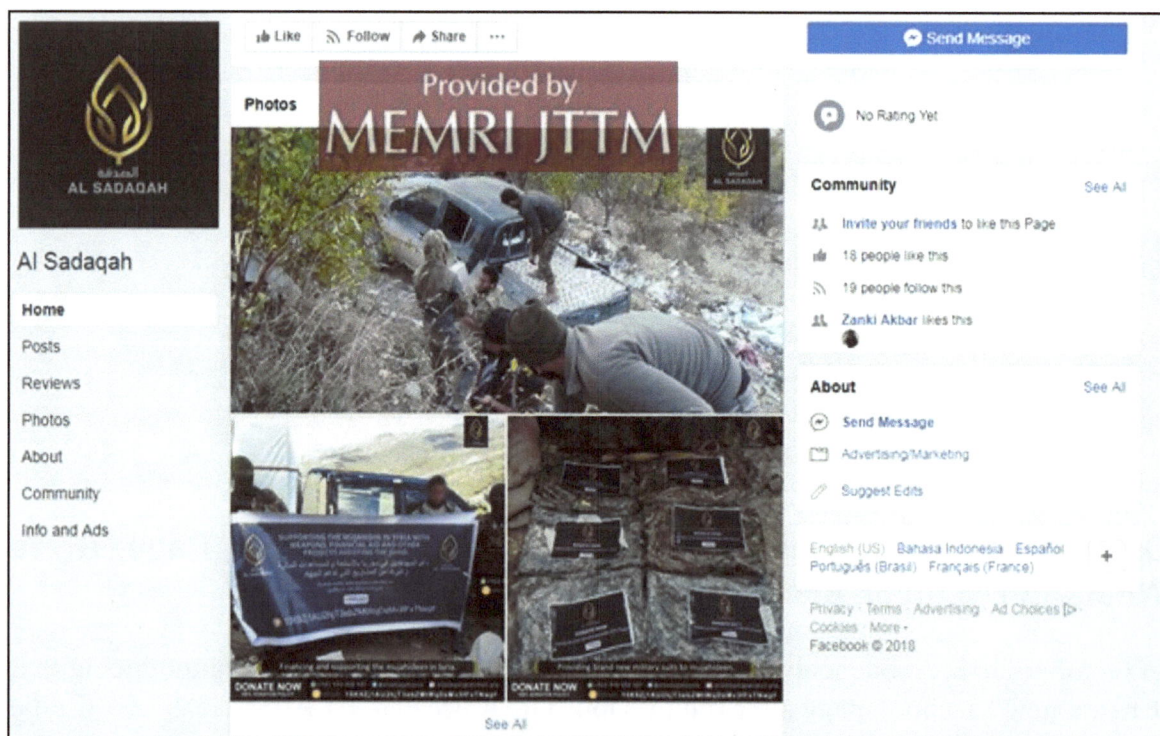

The Coming Storm: Terrorists Using Cryptocurrency

The profile picture for the Facebook page is the group's logo.

On October 29, Al-Sadaqah wrote on Facebook: "New channel after deletion. An independent charity organisation supporting the Mujhideen in Syria. Like and share."

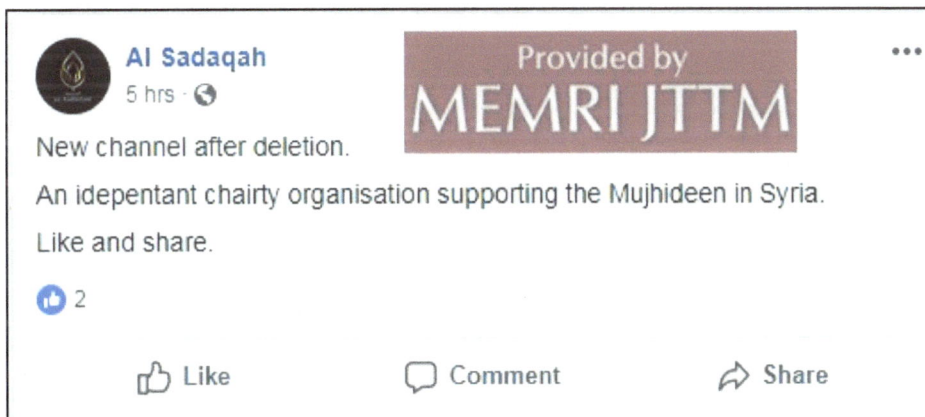

On October 29, Al-Sadaqah posted a photo of two men carrying a sign that reads: "Supporting the mujahidin in Syria with weapons, financial aid and other projects assisting the jihad." It also says: "Donate safely and securely with Bitcoin" and gives the Bitcoin address.

Stalinsky ■ M E M R I ■ ميمري

The bottom of the photo includes the group's information for WhatsApp, Telegram, and SecMail email address (alsadaqah@ secmail.pro).

The group also posted a photo of soldiers wearing new uniforms while holding packages bearing the Al-Sadaqah logo and Bitcoin address.

SadaqaCoins On Twitter And Telegram

SadaqaCoins operates both a Twitter and a Telegram account to promote its platform. On both accounts, an opening message announces the launch of the platform: "The Sadaqa-Coins Team is pleased to present a gift and service for the ummah. The accounts explain how to use the Orbot and Orfox apps for using TOR on an Android OS. The Orbot app for Android provides the core Tor service and connectivity to the Tor network for any app. The Orfox web browser for Android gives mobile phone users to secure communications through the Tor network.

The Coming Storm: Terrorists Using Cryptocurrency

Jihadi Groups Use Monero Cryptocurrency For Fundraising

Monero is a decentralized cryptocurrency created in 2014[85] that uses ring signatures,[86] confidential transactions, and stealth addresses[87] to obscure origins, amounts, and destinations of all transactions. It is a decentralized cryptocurrency, meaning it is secure digital cash operated by a network of users. Ring signatures enable the signing of transactions among a group of users so that outside observers cannot decipher the signer of a transaction.[88] Additionally, stealth addresses, or new addresses, are generated for one-time use to ensure only a sender and recipient has access to transaction details. Since all Monero transactions are untraceable, it is a secure, fast and private form of payment.[89]

Monero has been recognized for its "robust anonymity" compared to competitors due to its reduced traceability.[90] The cryptocurrency has been used by foreign governments such as North Korea to avoid international sanctions and by criminals participating in illegal or dark-web markets.[91]

Riccardo Spagni of South Africa is currently the lead developer of Monero. The location of Monero headquarters is not available online. Monero's Terms and Conditions do not explicitly mention terrorist or criminal use, though "Prohibited Conduct" includes: "Personal attacks on or abuse of any members, moderators, or administrators of the forum, postings for any unlawful or fraudulent purpose (including links), phishing, posting of offensive content including profanity, obscenity, racist or pornographic material and posting of any materials that are defamatory of infringe any person's rights."[92]

The following section will review examples of jihadi groups using Monero, such as the On the Ground News media outlet and the Indonesia-based Abu Ahmed Foundation, including some which have already been discussed in this report, such as SadaqaCoins and the Al-Sadaqah Organization. These bodies are, inter alia, raising funds for maintaining operations, obtaining food, weapons, and other equipment for jihadis in Syria, and for supporting fighters' families.

On The Ground News

The On The Ground News (OGN) media outlet, run by the pro-jihad American media activist Bilal Abdul Kareem who was close to Jabhat Al-Nusra (JN) and other jihad groups in Syria,[93] has asked its audience on Facebook and Twitter for funds via the Bitcoin and Monero cryptocurrencies. Bilal Abdul Kareem has on many occasions interviewed Al-Qaeda-linked militants in Syria and prominent jihadi figures such as Sheikh 'Abdallah Al-Muhaysini.[94] Kareem has filed a lawsuit in the U.S. challenging his alleged placement on a "kill list" by U.S. authorities in Syria, seeking to clear his name after what he claims were five near-misses by U.S. airstrikes there.[95]

Abu Ahmed Foundation (AAF)

The Indonesia-based Abu Ahmed Foundation (AAF) raises funds to buy food, weapons, clothing, and other materials for jihadis in Syria. It frequently posts, primarily in Indonesian, on its Facebook, Twitter, WhatsApp, and Instagram accounts. A graphic shared on

The Coming Storm: Terrorists Using Cryptocurrency

the AAF Media Centre Telegram channel promotes the AAF Facebook page and says that those interested can donate to a "*Ribath* [guard duty] Project" using cryptocurrencies such as Bitcoin, Monero, Dash, and Verge.[96]

Syria-Based Twitter Account Promotes Jihadi Groups, Raises Money Via Bitcoin And Monero

The @Aswed_F Twitter account, username "The Lattakian," which is run from inside Syria and started tweeting in 2015, updates its 6,788 followers on the activity of Hay'at Tahrir Al-Sham (HTS) and other rebel factions in the Idlib governorate of Syria. The account maintains bitcoin and Monero accounts, and links to content from "On the Ground News" (OGN) and includes links to the OGN page on the Patreon crowdfunding platform. It posts content critical of ISIS and hostile to the West. The profile reads: "Defend #Idlib Defeat #Bashar Defeat #Daesh – Zakat accepted to arbatash@tuta.io (including via Monero & Bitcoin.)" It also gives a link to the user's account on the cryptocurrency exchange platform Freewallet; it uses Freewallet's Monerowallet function, which converts incoming currency into Monero (XRM). The address is: 86U5dWizyVdMcZXe6ueo3F4H37kGkrePfa4dDX-UgobWFiFYgi4krF5gcFWYKE428n7PLcUsrc7KbhjJwdCcqUk4kKXBsRiq. In a January 2019 post, the account provided its bitcoin wallet address, 32WQFc982DJiY4EwpsVD5p-jpnD5G523BDJ; in another, it requested donations "to aid Sham [Syria]." In May 2019, it praised Incite the Believers operations room, an alliance of several militant groups with ties to Al-Qaeda, for having "massacred Assadists."[97]

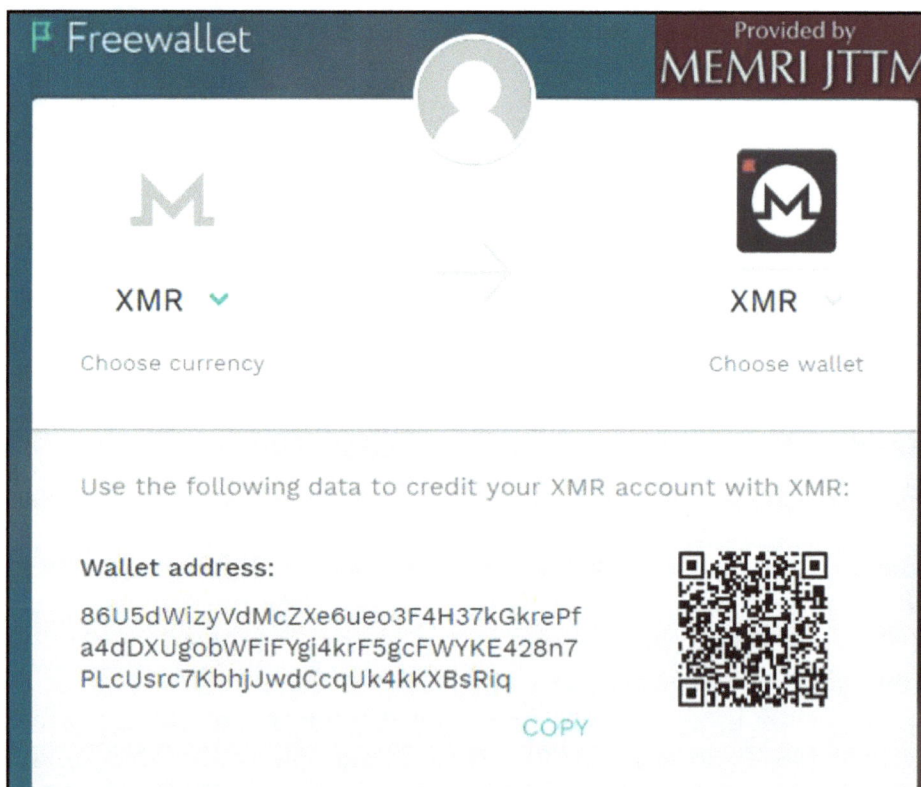

The Coming Storm: Terrorists Using Cryptocurrency

On January 16, 2019, the user asked for donations: "*ya ahl Sunnah* [Oh, Sunni Muslims]. Its never been cheaper, easier, safer to aid Sham. One paycheck today could potentially repel an aggression against the muslims. How abt thereafter. Ive done my part and cant benefit from *dunya* [the world] anymore, now their situation is on you. News will go nowhere w/ out effort."

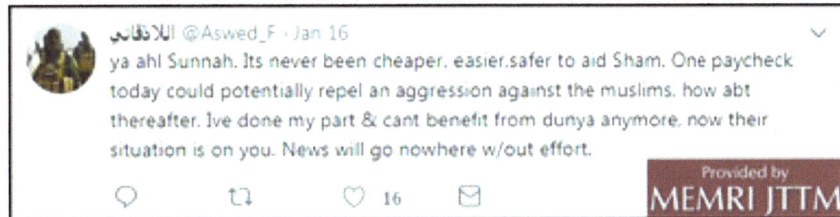

In a January 8, 2019 post, the user gave his Bitcoin wallet address. The address is: 32WQF-c982DJiY4EwpsVD5pjpnD5G523BDJ.

The user reposts other pro-jihadi journalists including Bilal Abdul Kareem's On the Ground News, retweeting links to OGN's Patreon page.

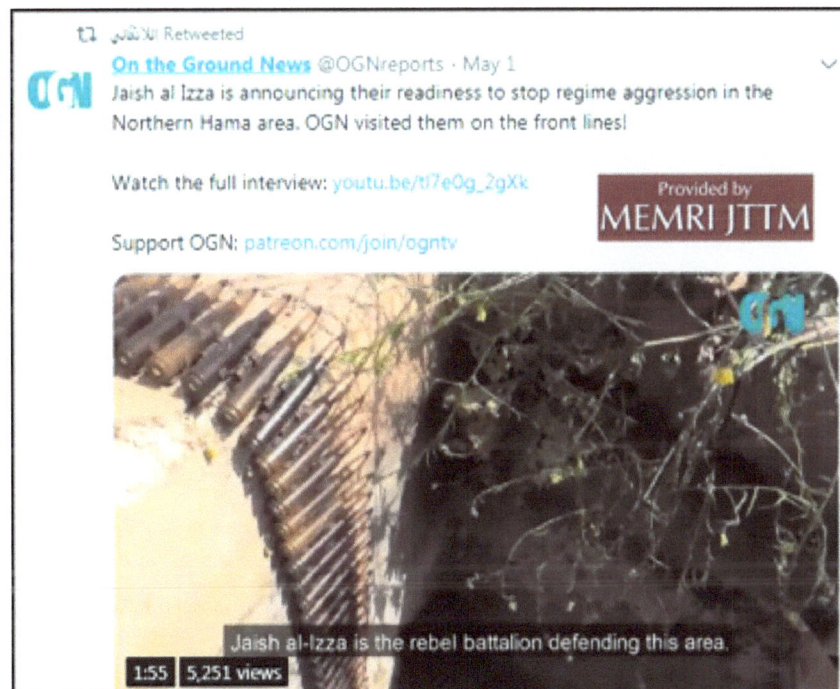

The Twitter account was linked to an account on the Curious Cat platform, where the user gave information about where he lives and what he thinks will happen to the Syrian revolution: "Free Lattakia. It's only the beginning, what started can't be stopped, regardless what armchair politicians and generals say." The Curious Cat platform has been used by pro-Al-Qaeda jihadis in Syria in the past.

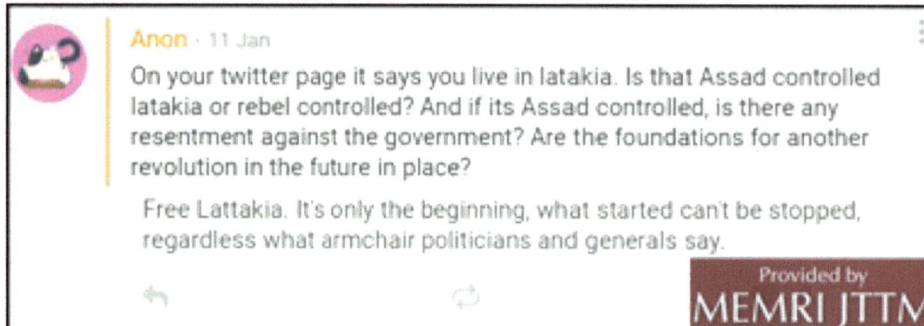

Anon · 11 Jan

On your twitter page it says you live in latakia. Is that Assad controlled latakia or rebel controlled? And if its Assad controlled, is there any resentment against the government? Are the foundations for another revolution in the future in place?

Free Lattakia. It's only the beginning, what started can't be stopped, regardless what armchair politicians and generals say.

Provided by
MEMRI JTTM

In an April 12, 2019 post, the user commented on the subject of the families of ISIS fighters: "Maybe send the kids to Idlib." In a subsequent tweet, he said: "We'll do everything for free including deradicalization for mothers, and there's even villages made for orphans. But it seems like [the coalition forces] just don't know how to kill em all without taking responsibility."

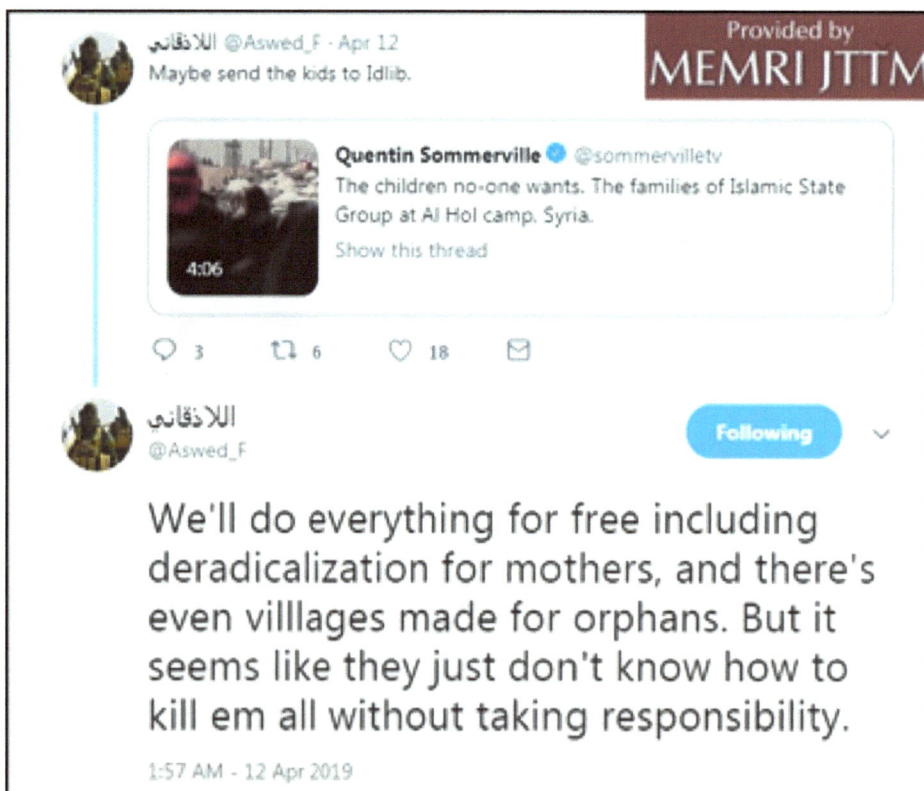

اللاذقاني @Aswed_F · Apr 12
Maybe send the kids to Idlib.

Provided by
MEMRI JTTM

Quentin Sommerville @sommervilletv
The children no-one wants. The families of Islamic State Group at Al Hol camp, Syria.
Show this thread

4:06

💬 3   🔁 6   ♡ 18   ✉

اللاذقاني
@Aswed_F

Following

We'll do everything for free including deradicalization for mothers, and there's even villlages made for orphans. But it seems like they just don't know how to kill em all without taking responsibility.

1:57 AM - 12 Apr 2019

The Coming Storm: Terrorists Using Cryptocurrency

A February 17, 2019 comment stated: "Quite clearly seen in the comments under here that westerners are often worse than the ISIS they oppose. In addition their own sources say more than 50 children have died under YPG & coalition controlled camps."

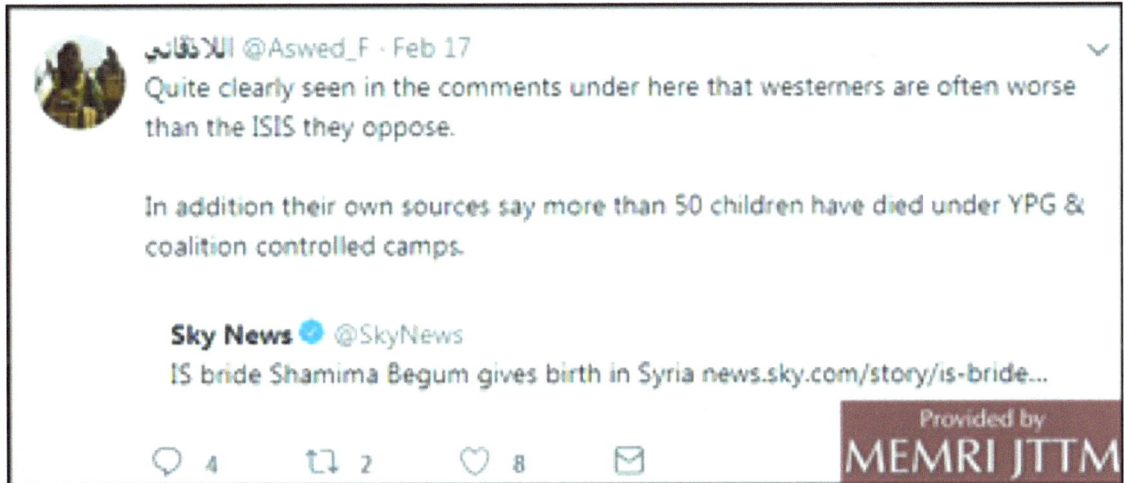

الذئب اللاذقاني @Aswed_F · Feb 17

Quite clearly seen in the comments under here that westerners are often worse than the ISIS they oppose.

In addition their own sources say more than 50 children have died under YPG & coalition controlled camps.

Sky News ✓ @SkyNews
IS bride Shamima Begum gives birth in Syria news.sky.com/story/is-bride...

💬 4   🔁 2   ♡ 8   ✉

A more typical post dated May 13, 2019 touted the recent accomplishments of Incite the Believers operations room, an alliance of several militant groups with ties to Al-Qaeda, Latakia, saying they "massacred Assadists."

الذئب اللاذقاني @Aswed_F · 7h
One of them seems so close, enough for a move with a blade, combat knife. Total domination.

💬 2   🔁 2   ♡ 17   ✉

Show this thread

الذئب اللاذقاني @Aswed_F · 7h
Incite the believers massacred Assadists in #Lattakia

Kamuflaj @kamuflaaaj
#Müminleri_teşvik_et operasyon odası Lazkiye kırsalındaki Raşo tepesinde rejim mevzilerine gerçekleştirdiği sızma görüntülerini servis etti.
Show this thread

💬 4   🔁 15   ♡ 45   ✉

SadaqaCoins

As noted, the SadaqaCoins crowdfunding platform, created to act as a secure connection between those who want to fund jihad and jihadis who need funding, maintains many on-going fundraising projects and allows donations through Bitcoin, Ethereum, and Monero. The Sadaqacoins website gives information on the Monero market and shares cryptocurrency's exchange rate for several other currencies.[98]

The Coming Storm: Terrorists Using Cryptocurrency

The Frequently Asked Questions section of the SadaqaCoins website reads: "Which secure payment methods does Sadaqacoins prefer?" The response is: "We recommend Bitcoins, as it is the most well-known, rapidly growing cryptocurrency which is easy to access. However, we prefer Monero because it is 100% anonymous, non-traceable and is safer than Bitcoin..."

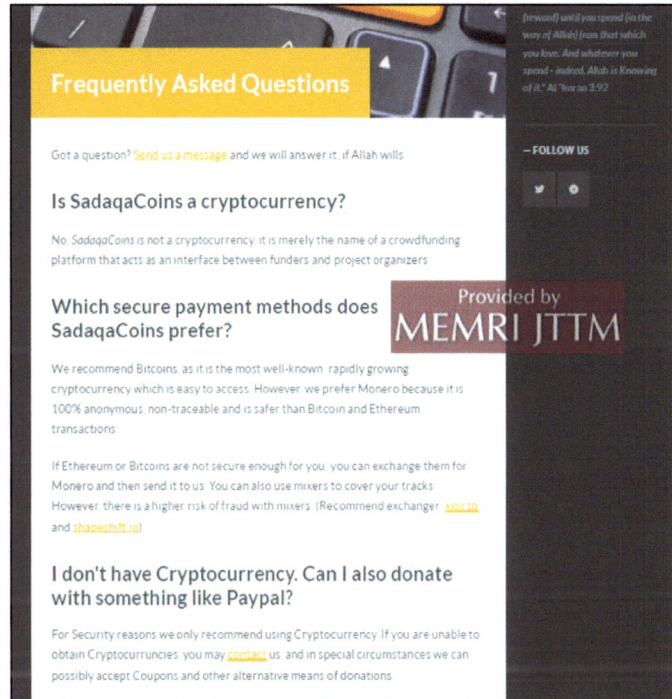

Sadaqacoins provides a link to donate using Monero and gives Monero as an option when listing the items for which the group is raising funds.

SadaqaCoins fundraises using Monero so jihadi groups can purchase items including:

- 4x4 pick up vehicle ($8,800)
- .50 cal bolt action sniper rifle ($4,070)
- .50 cal silencer ($550)
- Kestrel 4500NV "weather reading device"

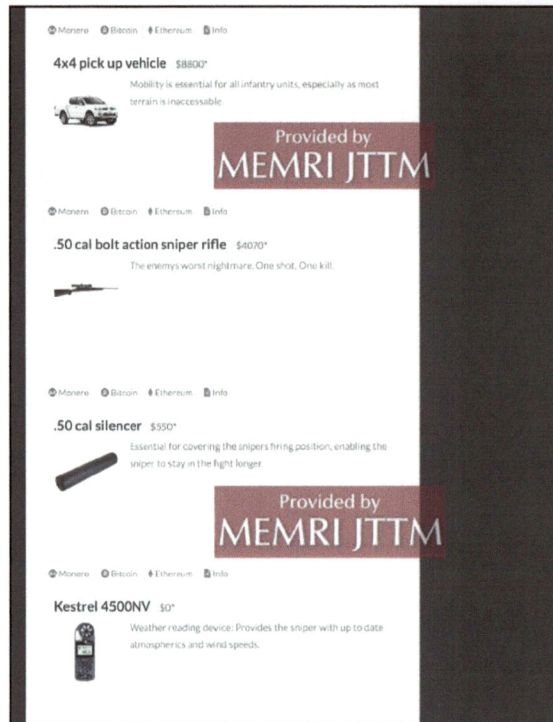

On January 3, 2019, the SadaqaCoins (New) Telegram channel posted the Monero payment ID JUNUDASHSHAM1 and a Monero wallet address.[99]

Al-Sadaqah Organization

Al-Sadaqah has been soliciting funds via cryptocurrencies including Monero to support the mujahideen in Syria for the past year. On its website's About page, Al-Sadaqah writes: "Our aim is to assist all mujahidin that are fighting and defending the front lines in Syria. Since 2011, the brave Mujahidin have been battling the Syrian regime in order to stop the oppression and injustice happening to our Muslim brothers and sisters. Our Muslim woman and children have been imprisoned, shot, killed and raped by this barbaric regime

The Coming Storm: Terrorists Using Cryptocurrency

and only a few chosen people of Ummah are defending their rights and our religion... The Jihad is Syria constantly needs your donations to survive and to continue the fight."[100]

Al-Sadaqah website donation page.

Donations can be sent using several cryptocurrencies listed on the website's Donate page, including Bitcoin, Dash, Monero, and Verge. Only the bitcoin wallet address is listed; users need to contact Al-Sadaqah for information on all other cryptocurrencies. The website also notes that the safest way to send money "without getting caught" is to use Dash or Monero.

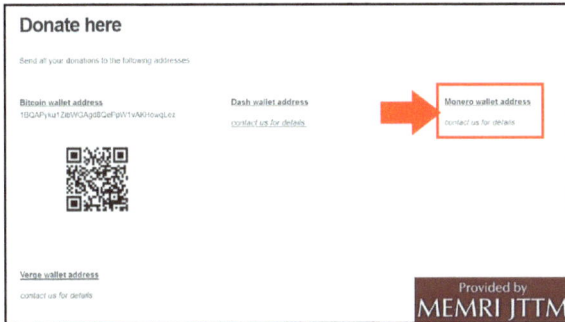

Al-Sadaqah social media accounts frequently share graphics encouraging supporters to donate using Monero, among other cryptocurrencies.

Earlier in 2018, Al-Sadaqah also shared a Monero address on its Twitter account. The group wrote: "Our temporary monero wallet address. Donate now to support the brothers on the front lines."[101] The post included the link to the Monero wallet and a barcode.

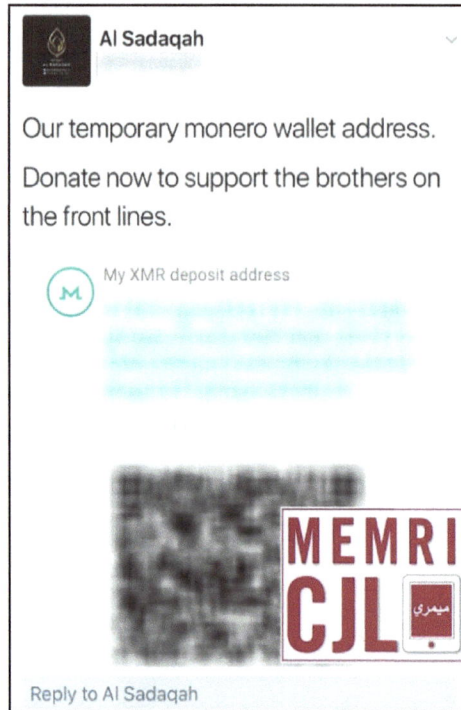

On February 23, 2018, Facebook user Musab Mujahid, whose profile picture was the Al-Sadaqah logo, posted a Monero deposit address and a bar code. He wrote on the post: "Our temporary monero wallet address...Donate now."[102]

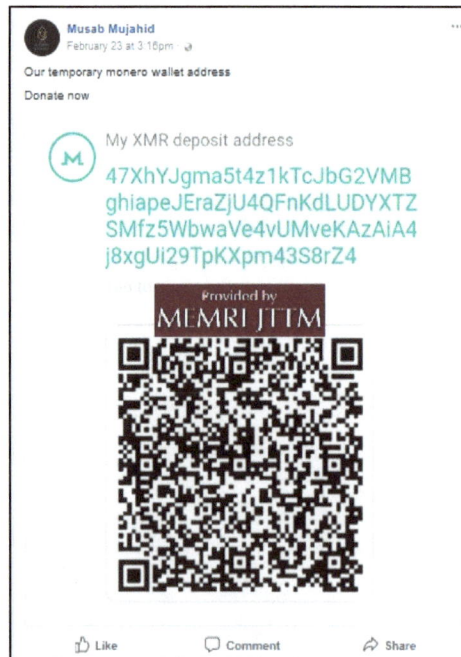

The Coming Storm: Terrorists Using Cryptocurrency

Leading Pro-ISIS Tech Group Electronic Horizons Foundation (EHF) Posts About Using Cryptocurrencies – Zcash, Bitcoin, Multiple Platforms – And Warns About Insecure Online Payment Services; Promises Future Articles On "Reliable" Cryptocurrencies

On December 11, 2018, the leading pro-ISIS tech group Electronic Horizons Foundation (EHF) published a list of questions and answers on a variety of technical topics.[103] EHF periodically posts on Telegram answers to questions it receives from users. The recent list includes some of the answers that EHF had posted in recent months.

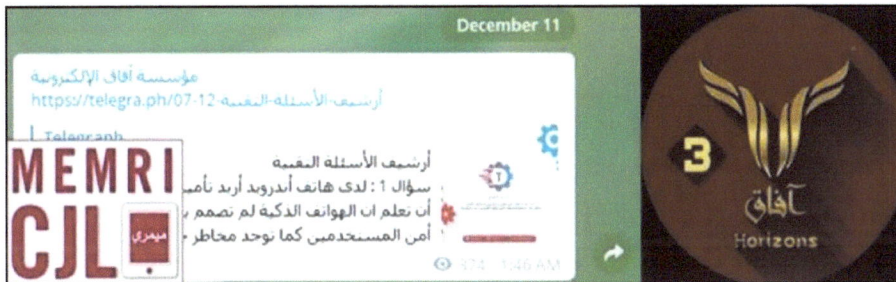

One question pertained to whether Bitcoin was "safe" to use for online purchases. EHF noted that although Bitcoin has a "high level" of secrecy and users need only the recipient's Bitcoin address to send funds, Bitcoin "does not provide a strong level of privacy," EHF wrote. The group recommended Zcash instead, noting its privacy features and the "difficulty in tracing it."

Zcash founder and CEO Zooko Wilcox said that the cryptocurrency "is a new blockchain and cryptocurrency which allows private transactions (and generally private data) in a public blockchain. This allows businesses, consumers, and new apps to control who gets to see the details

Post on Telegram: Bitcoin "does not provide a strong level of privacy."

of their transactions, even while using a global, permission-less blockchain." It is reportedly less transparent than bitcoin.[104] Launched on October 28, 2016, Zcash has a total supply of 21 million units.[105]

Stalinsky ■ M E M R I ميري

July 2019: Leading Pro-ISIS Electronic Horizons Foundation (EHF) Tech Group Warns Against Using Unsecure Online Payment Services, Promises Future Articles On "Reliable" Cryptocurrencies

On July 11, 2019, the pro-ISIS tech group Electronic Horizons Foundation (EHF) posted an article titled "An Introduction To Secure Financial Transactions," providing information about several online payment platforms and services and their security implications and warning of the security threats and challenges facing supporters of the mujahideen who send and receive money via the Internet. The article also warned about the risks involved in in-person meetings for giving or receiving money.

The Coming Storm: Terrorists Using Cryptocurrency

البطاقات البنكية (VISA / Mastercard / American Express)

ترتبط البطاقات البنكية بحساب بنكي يحتوي على بياناتك الشخصية وكل عملية شراء او تجريها او كل عملية تحويل اموال تخزن في سجلات البنك وبالطبع تتعاون البنوك مع الاجهزة الحكومية بإستخدام البطاقات البنكية في عمليات الشراء عبر شبكة الإنترنت كارثة أمنية بكل المقاييس وحتماً ستؤدي لاعتقالك بشكل او بآخر

خدمة الدفع الالكتروني Paypal

هي شركة أمريكية تتيح الدفع عبر بطاقات بنكية بدون اتاحة بيانات البطاقة البنكية للطرف الثالث في عملية الشراء مما يمنحك قليلاً من الخصوصية لكن جميع التعاملات المالية التي تقوم بها تخزن على سيرفرات الشركة سواء كانت عملية تحويل او سحب او شراء فضلاً عن عناوين الـIP التي استخدمتها في التسجيل او الدخول لحسابك وكذلك رقم الهاتف والبيانات الاخرى التي تطلبها الشركة كما ان باي بال تعرف من أنت وذلك في حال ربط البطاقة البنكية بحسابك والدعم من خلاله وبالطبع أجهزة الاستخبارات الامريكية لديها وصول مباشر لشركة باي بال

ويسترن يونيون Western Union

شركة أمريكية تقدم خدمات تحويل الاموال وبعض المتاجر والمواقع الالكترونية تعتمد عليها في عمليات الشراء لكن تطلب الشركة بطاقة هويتك في عملية التحويل وتحتوي افرع الشركة حول العالم على كاميرات مراقبة لتوثيق التعاملات المالية عند استلام او تحويل الاموال بذلك تعرف الشركة هويتك

التحويل البنكي

التحويل البنكي هو تحويل اموال لحساب بنكي معين لكن يضطر الطرف الذي يقوم بالتحويل الي كتابة استمارة تحتوي على بيانات الشخص المستقبل للحوالة المالية وكذلك بطاقة هويتك وتخزن تلك البيانات في سجلات البنك ونستطيع الحصول عليها الجهات الحكومية في اي وقت

المقابلة الشخصية

التعامل المباشر مع مناصر في عمليات تحويل الاموال من الاخطاء الشائعة التي يرتكبها المناصرين محتى وان كان من تتعامل معه ثقة فمى حال إعتقاله قد يدلي بمعلومات عنك مثل البيانات التي يعرفها طريقة التواصل ومكان الالتقاء واين تسكن وما مظهرك وجميع التفاصيل التي يعرفها عنك والتي قد تقودك للاعتقال ايضأ

بطاقات الدفع الفوري (One card) او (CashU) أو (Paysafe) ونظائرها الأخرى

بطاقات الدفع الفوري غالباً ما تحتوي على أكواد متسلسلة تدل على الموقع الجغرافي للمحل او المتجر الذي قمت بشراء البطاقة منه وربما الوقت والتاريخ في فاتورة الشراء ويحتفظ التاجر بتلك التفاصيل لكن اذا كان هناك كاميرا مراقبة في المتجر او المحل حينها تكون عملية الشراء غير آمنة لانه يمكن ربط الرقم التسلسلي للبطاقة او اكواد البطاقة مع ما قمت بشراءه عبر شبكة الانترنت ووقت وتاريخ الشراء مع سجلات تصوير كاميرا المراقبة حينها تُكشف هويتك

البتكوين (Bitcoin)

عملة بيتكوين هي عملة إلكترونية لا مركزية يتم تداولها عبر شبكة الإنترنت وتعتمد بشكل أساسي على مبادئ التشفير لذلك تعرف بـ"Cryptocurrency" أي عملة مشفرة ومن المعروف أن عملة Bitcoin تتمتع بقدر عال من السرية حيث أن كل ما تحتاجه لإرسال بعض البيتكوينات لشخص آخر هو عنوانه فقط لكن عمليات التحويل التي يتم تخزينها على البلوك تشاين (Block chain) وعلى الرغم من عدم معرفتك لهوية مالك أي عنوان فيمكن معرفة عدد البيتكوينات التي لديه والعناوين التي ارسلت إليه وتعقب عناوين الـIP من خلال المتاجر او الشركات التي قمت بالتعامل معها إذا عملة Bitcoin لا تقدم مستوى قوي من الخصوصية والامان لكن توجد انواع اخرى من العملات الالكترونية المشفرة يمكن الاعتماد عليها في عمليات تحويل الاموال او الشراء عبر شبكة الانترنت وسيتم توضحيها تباعاً في المقال القادم باذن الله

The article began by shedding light on key online payment platforms and their security, noting that "online financial transactions may be necessary in some cases, such as purchasing VPN services, or applications and programs used for media activity."

Listing platforms, the article mentioned "bank cards (Visa, Mastercard, American Express), PayPal, Western Union, bank transfers, personal encounters, prepaid cards (Paysafe, CashU, and One Card), and Bitcoin." It stated that bank cards are linked to personal data, so that there is a record of every transaction. Using cards to make online purchases, it stated, is a "security disaster," leading ultimately to arrest, because banks deal with government services.

Concerning PayPal, the article said that it is an American company that stores personal data and every transaction through linking it to the IP address, cautioning supporters that "American intelligence services have direct contact with PayPal company.'"

It added that Western Union is an American company that offers a money transfer service, warning that the company stores personal IDs and that its physical locations are equipped with surveillance cameras to document the sending and receiving of funds. It warned also

against conducting bank transactions from personal accounts, stating that this information is on record and can be accessed by government services at any time.

In-person meetings by ISIS supporters, it added, are "one of the most common mistakes made by supporters"; even if the person you are meeting is trustworthy, they may, if arrested, divulge personal information that would make it easy to identify you.

Regarding prepaid cards, it said, these are "unsecure," particularly if the location where you buy them is equipped with cameras, because you can easily be identified using the card number.

In closing, the article identified bitcoin as a decentralized digital currency that requires only the recipient's address, and added that transactions are recorded in a public record called a blockchain, allowing senders' IP addresses to be tracked. It concluded that "bitcoin does not offer a great degree of privacy and security," but that there are other "reliable cryptocurrencies" for transferring funds or making online purchases. These, it says, will be detailed in future articles.[106]

The Coming Storm: Terrorists Using Cryptocurrency

Malhama Tactical, A Russian-Speaking Jihadi Military Contractor Operating In Syria, Fundraises And Markets Merchandise In Indonesian Using Bitcoin To Support Its Training Camps, Buy Turkish Construction Materials

Another group that has been fundraising on Telegram in cryptocurrency is Malhama Tactical (MT). A jihadi private military contractor, MT, which operates in Syria, has worked with Jabhat Fath Al-Sham (formerly Jabhat Al-Nusra, the Al-Qaeda affiliate in Syria), as well as Ahrar Al-Sham. Created "around 2016," according to its current commander Abu Salman Belarus, it comprises about a dozen instructors, primarily former Russian military. Its members have participated in battles for Idlib, Aleppo, and other locations since at least 2015. It has helped train various forces in Syria, including from Hay'at Tahrir Al-Sham and from the Turkestan Islamic Party (TIP), including in the use of various weapons and equipment – its social media accounts show it training fighters with tanks, RPGs, and small arms, as well as in tactics and long-range marksmanship. It has a considerable social media presence; in November 2016 it recruited on Facebook for instructors who were prepared to "constantly engage, develop and learn."[107]

On January 29, 2019, MT released a short video highlighting the group's trainers' experience and skill in shooting, maneuvering, and using weapons such as grenades and RPGs. In the video, in which a jihadi nasheed plays in the background, the group shared the handles for its public and private Telegram channels, asked for financial support in Bitcoin, and shared a Bitcoin address and a QR code."[108]

The MT video gave viewers a Bitcoin address where they coul donate.

The video gave viewers a QR code.

The video showed fighters training using various weapons and tactics.

On October 18, 2018, the MT Telegram channel posted a message about a new project and fundraising campaign. According to the post, they "have created good conditions for training; practice grounds, lodging places and bathrooms." Now, they plan to modernize the training and are asking for funding for airsoft rifles. "One set of ball ammunition and charging costs about $600. To complete the work of the camp we need 16 sets." For this reason, MT is looking to raise $8,000. The message adds that 150 "brothers" undergo training at their camp each month. The post ends by giving another Telegram handle "for feedback": @TacticalReview. A similar appeal was circulated in Russian.

Malhama Tactical
As-salāmu ʿalaykum wa rahmatullāhi wa barakatuh dear brothers and sisters, this is an appeal to every Muslim and each of our subscribers and volunteers

We are starting a new project and launching a new fundraising campaign. This time the amount is large and the project is very important for our ummah - this is the preparation of our brothers against the enemies of Allah. We have created good conditions for training: practice grounds, lodging places, bathrooms, we have done all the necessary work on arranging praise be to Allah.

Now we want to modernize and update the ways of preparing our brothers. The best solution for this will be airsoft rifles. The advantages of this type of training are mass: the conditions of a real fight, testing and an objective assessment of readiness with further adjustment of all shortcomings, and as a result, an ideal preparation.
One set of ball ammunition and charging costs about $600. To complete the work of the camp we need 16 sets.

Everyone who can help with finance, please contact us, your reward will be on the Day of Judgment with Allah, who is not able to help with finance, please spread this information about the collection and make a repost. Do not forget us in your duas too.

The equipping of one NATO soldier costs about $ 10,000, and one Russian special forces soldier $ 7,000, millions are spent on their training. We are already far behind and we want to prepare our brothers using modern methods and teach them to shoot and conduct assault operations as efficiently as possible. By the grace of Allah, 150 brothers each month undergo training in our camp, doing various special courses and tactical exercises.

All other activists, please help and assist my brothers with this project. We can do it together, inshaAllah.

The total amount of $8000.
Jazakumulah janna, may Allah help us, receive from you and be pleased with you.

For feedback: @TacticalReview

Malhama Tactical.

397   edited 11:44 AM

The Coming Storm: Terrorists Using Cryptocurrency

СТРАЙКБОЛЬНЫЕ ПРИВОДЫ
КОМПАНИЯ ДЛЯ СБОРА СРЕДСТВ

Для еще более эффективной и безопасной подготовки наших братьев необходимо **16** комплектов страйкбольных приводов

1 привод со всеми необходимыми аксессуарами стоит **600$**

Provided by
MEMRI JTTM

Для связи | *SalmanBelarus*

MALHAMA TACTICAL

MT raises funds on social media, requesting donations in Bitcoin and by other means. On August 4, Abu Salman tweeted: "Dear brothers, our project needs help and support, you can donate for our team by Bitcoin wallet anonymously and safely, text me DM. Please retweet and share."

Abu Salman Belarus @SalmanBelarus · 4.08

Dear brothers, our project needs help and support, you can donate for our team by Bitcoin wallet anonymously and safely, text me DM. Please retweet and share.

Malhama Tactical.

Provided by
MEMRI JTTM

♡ 1   ↻ 7   ♡ 22

On October 8, 2018, Abu Salman tweeted a video showing sand, cinder blocks, and bags of cement marked with the name KÇŞ Çimento, a cement company based in Kahramanmaraş, Turkey,[109] and noted that they were obtained with "the help coming from our brothers" but that the need for "support" was ongoing. He wrote in Turkish: "Praise be to Allah, with the help coming from our brothers we got some material for the establishment of a training camp. The project is continuing but we are still missing some things, those who want to give support and help can send [me] a message."

Twitter.com/SalmanBelarus/status/1049296540744015872

The Coming Storm: Terrorists Using Cryptocurrency

The group also appears to sell tee shirts with its logo. On July 17, 2018, Abu Salman tweeted photos of shirts along with color and size information, with text in Russian.

Abu Salman Belarus @SalmanBelarus · 17.07

В восточной Азии продаются футболки с нашим лого, фаны)

Malhama Tactical.

☐ 6   ⇄ 1   ♡ 18

Abu Salman Belarus @SalmanBelarus · 17.07

Скоро начнутся курсы по подготовке профессиональных стрелков ПКМ от команды инструкторов МТ, подготовка будет проходить в течение 10 дней, запишитесь за ранее. Подробнее можете узнать на нашем канале в телеграм:

Malhama Tactical
Go ahead, press the link MT

t.me

☐   ⇄ 3   ♡ 9

Twitter.com/SalmanBelarus/status/1019286393363853313

MT tee shirts are also offered for sale by others. For example, a July 29, 2017 Instagram post by @balaruna_attack featured photos of an MT tee shirt; the text, in Indonesian, asked "Who likes to play airsoftgun" and described the product. Additionally, a November 10,

2017 Instagram post by @zinkistore showed an MT tee shirt, writing "Collection of @zinkistore da'wah shirts. Suitable for young people today. For leisure or certain events." The English in the post read: "Ready stock!"

Instagram.com/p/BXHnw3LFy3t/?tagged=ak47lovers; Instagram.com/p/BXHnq6CF5pp/?tagged=ak47lovers

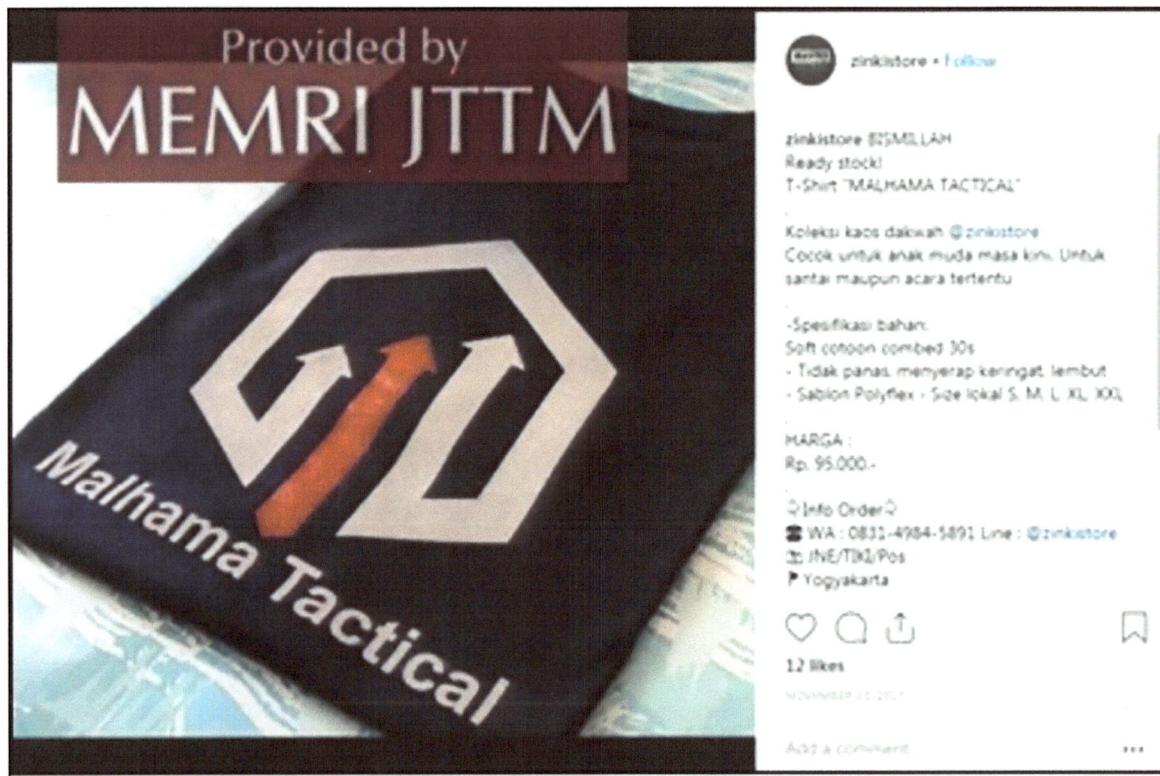

Instagram.com/p/BbUM-EJFmAD/?taken-by=zinkistore

The Coming Storm: Terrorists Using Cryptocurrency

On November 11, Abu Salman tweeted: "Urgent! Our brothers in training camp need solar panel or generator and battery. Please help with purchase, it costs almost $200. You can help anonymously and safely, please text me DM."

Abu Salman Belarus @SalmanBelarus · 11.08

Urgent! Our brothers in training camp need solar panel or generator and battery. please help with the purchase. it costs almost $200. You can help anonymously and safely. please text me DM.

Malhama Tactical.

Provided by
MEMRI JTTM

♡ 9   ♡ 11

Al Ansaar – Aid For Syria, Jihadi "Charity Organization" On Telegram, Fundraises For Mujahideen And Foreign Fighters In Syria – Including Australians, Swedes, Other Europeans – And For Purchasing Drones And Other Equipment, Via Telegram And WhatsApp And Using Bitcoin Address

Al Ansaar – Aid For Syria, which bills itself as a "charity organization," maintains a Telegram channel that it uses primarily for raising funds for the mujahideen in Syria. The group uses Bitcoin addresses, Telegram, and WhatsApp to collect money to supply fighters with equipment such as weapons and drones, as well as food and clothing for fighters and their families. As of this writing, the Telegram channel has 129 members, 42 photos, 10 videos, and four audio files; its first post was on July 2, 2018. The channel's description states: "This is a completely non profit, independent charity organization and we are not affiliated with any particular groups."

The following section focuses on the Telegram activity of Al Ansaar – Aid For Syria, including promoting and soliciting for the group's numerous projects; soliciting donations; messaging, posts of photos and information on the cost of supplies it purchases, and repeatedly posting contact information and its Bitcoin address for donations.

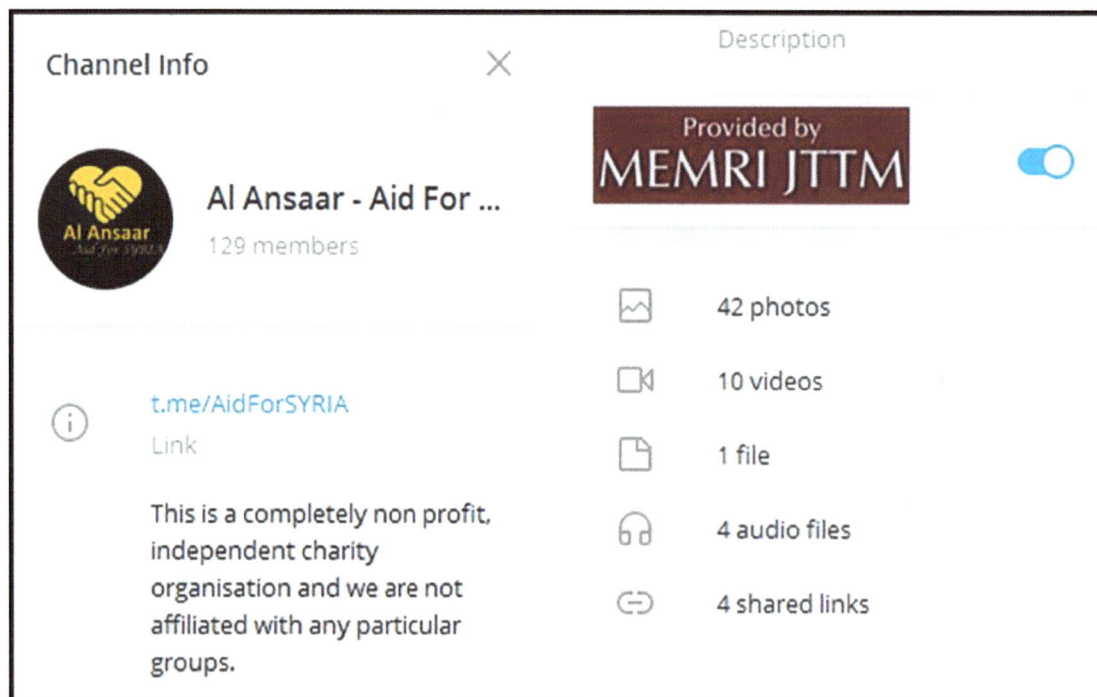

Channel information for Al Ansaar – Aid For Syria

The Coming Storm: Terrorists Using Cryptocurrency

Al Ansaar's Mission And Goals

Al Ansaar's first post on its Telegram channel, on July 2, 2018, stated: "We have been fortunate enough to be blessed with *hijrah* to the lands of *shaam* [Syria] and become involved in some projects assisting those who give their lives to support this ummah." The message went on to say that the channel will be used as an outlet for "encouraging people to spend in Allaahs way" and "to offer anonymous means for those outside shaam who desire a share of this great reqard... The virtues and rewards of spending on jihaad in Allaahs way are numerous and cannot be achieved in any other form of good except jihaad in Allaahs way or assisting jihaad financially... Please keep us in your *duas* [supplications] and spread this channel far and wide remembering that any good that comes from someone you share it to will be in your account as well as theirs on the day of *qiyaamah* [resurrection, i.e. day of judgment]."

> **Al Ansaar - Aid For SYRIA**
>
> Slms beloved Brothers, we have been fortunate enough to be blessed with hijrah to the lands of shaam and become involved in some projects assisting those who give their lives to support this ummah.
>
> We had not posted much media before but insha Allaah will use this channel as an outlet for the media of works past and present in the aim of encouraging people to spend in Allaahs way and also to offer anonymous means for those outside shaam who desire a share of. this great reward, and who desire to assist the defenders of Islam.
>
> The virtues and rewards of spending on jihaad in Allaahs way are numerous and cannot be achieved in any other form of good except jihaad in Allaahs way or assisting jihaad financially.
>
> Zaid bin Khalid (May Allah be pleased with him) reported:
>
> The Messenger of Allah (ﷺ) said, "He who equips a Ghazi (fighter) in the way of Allah is as if he has taken part in the fighting himself; and he who looks after the dependants of a Ghazi in his absence, is as if he has taken part in the fighting himself."
>
> [Al-Bukhari and Muslim]
>
> **Provided by MEMRI JTTM**
>
> Please keep us in your duas and spread this channel far and wide remembering that any good that comes from someone you share it to will be in your account as well as theirs on the day of qiyaamah.
>
> May Allaah reward us and you all with jannah firdous without hisaab and make the way there full of aafiyah for us all.
>
> 👁 84   edited 6:49 AM

On July 8, 2018, Al Ansaar wrote, under the heading "Priorities": "A group of criminals break into your parents house killing and torturing anything in their way. What is the priority? A) wait for those inside to flee and then if any make it give them shelter. B) sit and make dua with no action. C) Go in and stop those who are causing the problem." The message concluded: "Its still not too late. DONATE NOW: Telegram contact: @AidForSYRIA1, Bitcoin [Wallet address] 1EutyMiNCEosn4xxAM9Avvir6fyWwUjdi, Plz Donate and spread the channel @AidForSYRIA."

Al Ansaar - Aid For SYRIA
Priorities

I want to give u a scenario, for those with understanding.

A group of criminals break into your parents house killing and torturing anything in their way.

What is the priority?
a) wait for those inside to flee and then if any make it give them shelter.
b) sit and make dua with no action
c) go in and stop those who are causing the problem.

Its brainless question I know?
Then y is it that we are so unwilling to support the mujaahideen with our lives and wealth but we see so many wanting to spend on aid. No matter how many we aid, if we don't stop the problem the numbers will only increase right.

And all is good, no doubt, but the least we can do is be honest with ourselves and change 4 the better.

Its still not too late.

Provided by
MEMRI JTTM

DONATE NOW:
Telegram contact: @AidForSYRIA1

Bitcoin: 1EutyMiNCEosn4xxAM9Avvir6fyWwUjdi

Plz Donate and spread the channel.
@AidForSYRIA

👁 99   edited 9:41 AM

The Coming Storm: Terrorists Using Cryptocurrency

Al Ansaar's Drone Project

On July 2, Al Ansaar posted a photo of a rifle and a box for a DJI Phantom 4 Pro drone with a camera. The text below the image read: "Providing mujaahideen with a drone for reconnaissance purposes. Please assist us by sharing and donating."

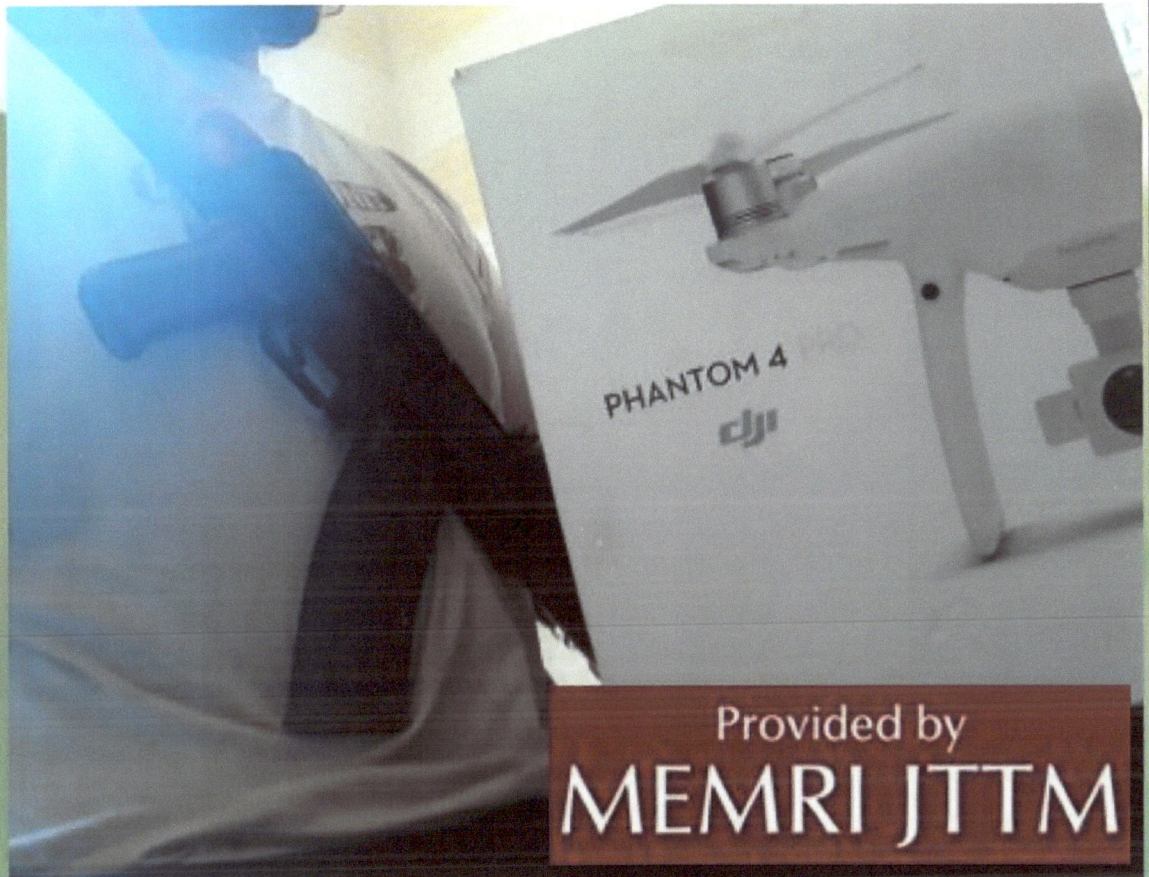

Al Ansaar - Aid For SYRIA

PHANTOM 4
dji

Provided by
MEMRI JTTM

Providing mujaahideen with a drone for reconnaissance purposes.

Please assist us by sharing and donating.

👁 81   8:30 AM

One post focused on purchasing drones as part of the group's "Equipping Brothers" project: "Providing drones to brothers so that they may scout the enemy territories without the risk of brothers lives before the missions. These drones although small are very powerful and extremely helpful to the mujaahideen. Drones cost between 1500 – 2500 $."

Al Ansaar - Aid For SYRIA
4th project Equipping Brothers

Providing equipment to brothers:
Recently some good opportunities have come up for the mujaahideen to inflict damage on the kuffaar, and part of preperation is to get equiped with some decent equipment even though we trust Allaah only.

- drone project
Providing drones to brothers so that they may scout the enemy territories without the risk of brothers lives before the missions.

These drones although small are very powerful and extremely helpful to the mujaahideen.

Drones cost between 1500 - 2500 $

- special force tactical equipment
The vision is to get some equipment that is especially helpful in night raids
For a small group between 6 and 10.

This equipment is really costly but the benefits are huge, the idea is that we have this equipment with us and that any trusted groups we know who have good plans for an attack we lend them this specialised equipment for their mission.

The Coming Storm: Terrorists Using Cryptocurrency

Other Projects

"Assisting Brothers" – Including From UK, Sweden, Australia, And Malaysia

On August 31, 2018, the group wrote about its "Assisting Brothers Project," which included "some messages from the brothers thanking donors for the donations." The post mentioned British, Swedish, Australian, and Malaysian "brothers." It included a section titled "Donations or Information: Telegram contact @AidForSYRIA1, Bitcoin [Wallet address] 1EutyMiNCEosn4xxAM9Avvir6fyWwUjdi" and added "Plz Donate and spread the channel @AidForSYRIA."

Al Ansaar · Aid For SYRIA
ASSISTING BROTHERS PROJECT

A major portion of our support comes from a SISTER, yes a SISTER far away who has been assisting us kindly for some time. below are some messages from the brothers thanking donors for the donations.

British bro:
Appreciation to the sister supporting the mujahideen.

The Messenger and those who believe with him, strive hard and fight with their wealth and lives in Allah's Cause.

All of the ayat regarding jihad begin with striving with wealth before lives except one. "Indeed, Allah has purchased from the believers their lives and their properties [in exchange] for that they will have Paradise."

Surely we the mujahideen consider spending in the fisabillah the best deed along with fighting as this is a cause of supporting mujahideen, with equiptment, food, drink, medicine and feeding families.

So you may see it as something little as you don't fully comprehend the effects your spending has on the mujahideen but its big to Allah. We respect you very much for your courage and we ask Allah to give you jannah firdaws ala and tenfold in that which you have spent.

The parable of those who spend their substance in the way of Allah is that of a grain of corn: it groweth seven ears, and each ear Hath a hundred grains. Allah giveth manifold increase to whom He pleaseth: And Allah careth for all and He knoweth all things. (Surah Al-Baqara, 261)

Swedish bro
Salam alaykum

May Allah give you the highest palaces of jannah sister, your help for us is to much to describe in words only Allah can pay you back. I hope you and we are from the people who dua is answered. i ask Allah to increase you in his blessings and accept from you the good deeds you put forth and forgive your errors. Dont forget us in your supplications 🙌

Australia bro
Say to her may allah reward her and give her the ajr of jihad and tell her hadith of who eases a muslim in dunya allah will ease their way inshallah in akhira and tell her how much help she is actually giving is a lot of help to the bros. And money is a major factor that if its not available the whole jihad is not available

Malaysian bro
M M2:
Tell her that Its a big thing for a brother who have left everything he had for the sake of Allah to know that there is still someone out there who still cares about them. When even the people amongst us here would betray them in a hearbeat. May Allah unite all of us in the best gathering in Jannahtul firdaus al a'la

DONATIONS OR INFORMATION:
Telegram contact: @AidForSYRIA1

Bitcoin: 1EutyMiNCEosn4xxAM9Avvir6fyWwUjdi

Plz Donate and spread the channel.
@AidForSYRIA

👁 80 9:28 AM

Subsequently, July 9, 2018 posts featured a photo of an Al Ansaar sign and items apparently purchased with donations, among them a solar panel and an automatic weapon, along with blankets and carpets. The accompanying text noted, under the headline "Assisting brothers project," that these items were part of "helping a newly married brother with necessities for his house. Jihaad is extremely taxing mentally, but by the mercy of Allahh, He gets brothers married to amazing siters who Allaah blesses with the imaan to marry someone who could be killed at any moment. Some think jihaad is 4 or 5 days and then you will be shaheed. It is a long path and many brothers marry to assist them with being firm on this path." It added: "DONATE NOW: Telegram contact: @AidForSYRIA1, Bitcoin [Wallet address] 1EutyMiNCEosn4xxAM9Avvir6fyWwUjdi, Plz Donate and spread the channel @AidForSYRIA."

"Bros In Need" – Foreign Fighters

A July 4, 2018 post set out some of the group's ongoing projects. The first project, "bros in need" emphasized that it was aimed at fighters "from diff places who came to help the suffering ummah for Allaahs sake, they are each going through different difficulties BY CHOICE FOR ALLAAHS SAKE. Most could leave and earn big money in their home lands but they have put the call of Allaah above everything." It added: "I spoke to a sheikh and he says the best policy with zakan is to give cash to the people, because they know what they need most, he said these mass iftaar and food packs make for good media but the benefit lies in seeing to the peoples needs best. Cost of supporting average family is 100$ per month."

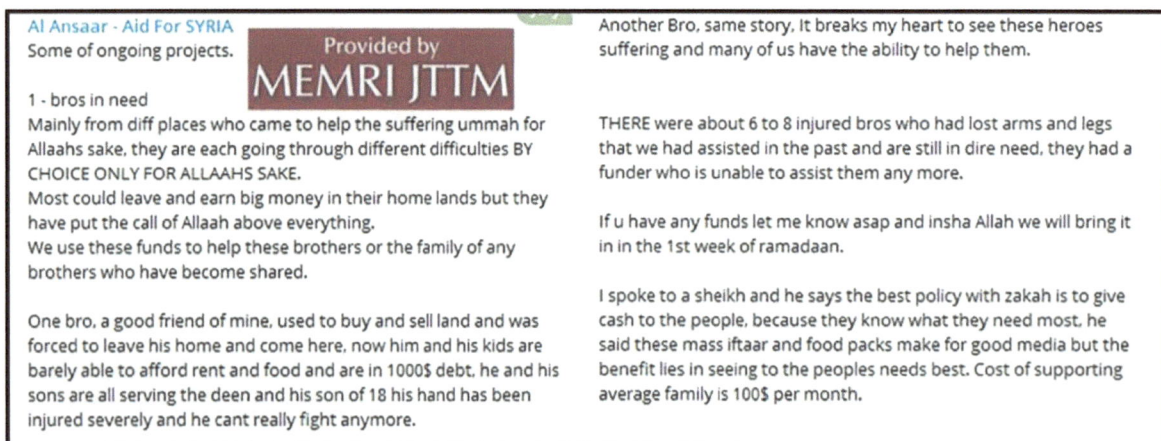

The Coming Storm: Terrorists Using Cryptocurrency

"Sisters In Prison"

2 - sisters in prison

Provided by
MEMRI JTTM

Alhamdulillah from the donations received 2 sisters and a brother have been taken out of prison.

These are lives of families that have suffered greatly that Allaah has allowed us to help change for the better. What an honour it is to be allowed to do this.

The situation is very sad and at times unbelievably that we leave them to rot without doing anything and this is 1 of the FARDH in Islam.

We need to spread the awareness then even if we ourselourselvs can't give anything, if some1 else does we get the full reward without decreasing their reward.
How merciful Allaah is.

I have a video soon of 1 of the leading muhaddithin in sham speaking and encouraging for us to help our sisters and brothers which I will translate and send out soon.

A post about a second project, "sisters in prison," stated: "Alhamdulillah from the donations received 2 sisters and a brother have been taken out of prison. These are lives of families that have suffered greatly that Allaah has allowed us to help change for the better. What an honour it is to be allowed to do this." The message also said: "I have a video soon of 1 of the leading *muhaddithin* [traditionists i.e., hadith scholars] in sham speaking and encouraging for us to help our sisters and brothers which I will translate and send out soon."

An update on the "Freeing Sisters" project set out how donations had freed some women: "Now Allaah has honoured us with hijrah and made up part of freeing them from the prisons of those who would defile them." A "breakdown of our involvement until this moment" states: "The payments from Al ansaar – Aid 4 syria for the captives. 4500\$ +1800 +6883 +600=14283...2000\$ paid for the sister Manal Khalid Suriem...(Freed) 1200\$ for the brother Ibrahim Al-Tadmuri...(Freed) 3000\$ for the sister Mayada Abdulkarim Al-Haw in Madaya, Syria...(Freed) 570\$ For freeing the sister Mona from the air force intelligence section in Damascus in (Al-Tal) area. 1500\$ for freeing the sister Iman Hassan Benienen..(Damascus)."

Al Ansaar - Aid For SYRIA
FREEING SISTERS PROJECT

Provided by
MEMRI JTTM

It makes me cry from joy at times to be part of this project. I would sit in my country of birth and see the filthy treatment of the kuffar on our precious sisters, now Allaah has honoured us with hijrah and made us part of freeing them from the prisons of those who would defile them.

I speak to all muslim sisters out there, we would give our life instantly to defend the honour of any one of you, please remember us in duas and also assist this project by donating and sharing our channel.

This below is a breakdown of our involvement until this moment. May Allaah accept.

Bissmillah

The payments from Al ansaar - Aid 4 syria for the captives.

4500\$+1800+6883+600=14283

2000\$ paid for the sister Manal Khalid Suriem... (Freed)

1200\$ for the brother Ibrahim Al-Tadmuri.... (Freed)

3000\$ for the sister Mayada Abdulkarim Al-Haw in Madaya, Syria .... (Freed)

570\$ For freeing the sister Mona from the air force intelligence section in Damascus in (Al-Tal) area.

1500\$ for freeing the sister Iman Hassan Benienen.. (Damascus)

A July 29, 2018 update on women freed from prison included a message from one of them. The post concluded with: "Donations or Information: Telegram contact @AidForSYRIA1, Bitcoin [Wallet address] 1EutyMiNCEosn4xxAM9Avvir6fyWwUjdi, Plz Donate and spread the channel @AidForSYRIA."

"House Of Brothers"

Project No. 3, "house of brothers," was described as follows: "Alhamdulillah there is a house of brothers from all over the world that we have been supporting, some1 has sponsored their iftaar for ramadaan Alhamdulillah, and there are many other expenses monthly that need to be fulfilled." The post included a message from "Abdul Hameed from Europe" thanking Al Ansaar.

"Equipping Brothers"

A post about a fourth project, "Equipping Brothers," noted numerous types and prices of equipment required by fighters in Syria and requests donations for purchases, specifically for a "drone project" so that "brothers may scout the enemy territories without the risk of broth-

The Coming Storm: Terrorists Using Cryptocurrency

ers lives before the missions." Drones, it added, "cost between 1,500 - 2,500 $." It continued with a request for funds for equipment: a "suit not seen by thermals" for $300, a silencer for $150, "night vision" for $350, a "rail and holder" for $80, a helmet for $50, and body armor for $200. The post also noted that there is a need for "thermals," which are "game changers," and that a good one "is around 6,000$." It estimated "cost per guy" at "+-$1100."

Al Ansaar - Aid For SYRIA
4th project Equipping Brothers

Providing equipment to brothers:
Recently some good opportunities have come up for the mujaahideen to inflict damage on the kuffaar, and part of preperation is to get equiped with some decent equipment even though we trust Allaah only.

- drone project
Providing drones to brothers so that they may scout the enemy territories without the risk of brothers lives before the missions.

These drones although small are very powerful and extremely helpful to the mujaahideen.

Drones cost between 1500 - 2500 $

Provided by MEMRI JTTM

- special force tactical equipment
The vision is to get some equipment that is especially helpful in night raids
For a small group between 6 and 10.

This equipment is really costly but the benefits are huge, the idea is that we have this equipment with us and that any trusted groups we know who have good plans for an attack we lend them this specialised equipment for their mission.

An example of 1 man equiped:

Suit not seen by thermals 300$
Silencer 150$
Night vision 350$
Rail and holder 80$
Helmet 50$
Body armour 200$

ESTIMATES COST PER GUY : +- $1100

Provided by MEMRI JTTM

- Thermals

Thermals are game changers, anyone who has used them knows the massive difference they make, night becomes like day, but of course there is a price to pay.

A good thermal is around 6000$ and we are in need of them.

Anyone who wants to even donate a small amount towards this should know that the guy with the thermal usually gets many kills so you will get this and also insha Allaah.

And Allaah doesn't join the muslim and the kafir he killed in naar, so it's a freedom from jahannam card to kill the enemies of Allaah. How fortunate to be a part of this

👁 99   edited 4:59 AM

"Project Ribaat"

Al Ansaar - Aid For SYRIA
5th project Ribaat

Provided by MEMRI JTTM

We were blessed in ramadaan to assist the brothers in ribaat providing them with their food costs for the time they spent out in ribaat.

The cost per person per day in ribaat is minimal and the reward is amazing so get your share.

Plz share our link
@alansaarshaam

👁 104   5:47 AM

A post on a fifth project, "Ribaat [guard duty]," stated: "We were blessed in ramadaan to assist the brothers in ribaat providing them with their food costs for the time they spent out in ribaat. The cost per person per day in ribaat is minimal and the reward is amazing so get your share."

On July 2, Al Ansaar posted photos of parcels of food, drinks, and weapons described as "Ramadaan distributions of food to the ribaat [guard duty] points."

Al Ansaar - Aid 4 syria

Al Ansaar - Aid 4 syria
Photo

Provided by MEMRI JTTM

Assisting brothers Project

Our latest distributions starting as the winter period kicks in and those with kids especially require additional funds for keeping warm.

Our distributions go to brothers who don't have sufficient income to live minimal lives, yet in the blink of an eye they will leave their homes for Allaahs sake, what a contrast to those in the lap of luxury refusing to spare anything in Allaahs way to support these heroes.

DONATIONS OR INFORMATION:
Telegram contact: @AidForSYRIA1

This is the NEW BITCOIN ADDRESS.

1Dwnb1poPbu3XHNEoZA7VE6KnSAbgm4XEU

Plz Donate and spread the channel.
https://t.me/joinchat/AAAAAFJOW6S3THXCUQgz4Q

Telegram
Al Ansaar - Aid 4 syria
This is a completely non profit, independent charity organisation and we are not affiliated with any particular groups.

👁 106   10:31 AM

The Coming Storm: Terrorists Using Cryptocurrency

ALHAMDULILLAH, what did we expect and come here for?

So may be 1 of the biggest and most decisive battles in shaam coming up, we ask Allaah to give us thabaat and make it easy for us make us a great part of it and to give victory to the believers and humiliate the kuffaar and cast terror in their hearts and destroy them by our hands.

To those not here this is a very unsure time in shaam and things are changing so those who appreciated the fact that Allaah azza wa jam ONLY TGROUGH HIS MERCY opened a door for us to spend and assist these amazing causes now is the time to take most advantage of it.

If victory comes and all the Muslims in the world join and cheer insha Allah and they spend much at that time, it will not be the same as little spent in the toughest times.

If we are defeated and these doors of good are closed to you all you may never get another opportunity like this to spend in jihaad in His way.

To put this into perspective:
A poor man who spends 1000 on jihaad gets at lest the reward of spending 700 000.

A rich man who is not blessed with spending in this path has to spend 70 000 in other avenues to at least get reward of spending 700 000.

So NOW is the time to make our PROFIT.

Lastly we ask you all to forgive us any shortcomings and remember us all in your duas and to make the jihaad of the true mujaahideen your jihaad because they fight for your religion.

We still have the following causes ongoing:
1 - FREEING SISTERS from shiah prisons.
2 - Equipping brothers
3 - Funding needy brothers living costs.
4 - Fuding a house off brothers from all over the world.

NEW CAUSE EID UL ADHA
We intend to sacrifice at least 1 cow and distributing insha Allaah, a share will cost from 150 to 200 $.

DONATE NOW:
Telegram contact: @AidForSYRIA1

Bitcoin: 1EutyMiNCEosn4xxAM9Avvir6fyWwUjdi

Plz Donate and spread the channel.
@AidForSYRIA

⊙ 77   8:47 AM

Raising Funds To Fight Assad Regime, Russian Forces: "A Poor Man Who Spends 1000 On Jihaad Gets At Least The Reward Of Spending 700 000"

On July 30, 2018, Al Ansaar posted a message asking followers "to forgive us any shortcomings and remember us all in your duas and to make the jihaad of the true mujaahideen your jihaad because they fight for your religion." The message also included a summary of the ongoing causes: "freeing sisters from shiah prisons, equipping brothers, funding needy brothers living costs [and] funding a house off brothers from all over the world." It added: "To put this in perspective: A poor man who spends 1000 on jihaad gets at least the reward of spending 700 000" and concluded: "We intend to sacrifice at least 1 cow and distributing insha Allah, a share will cost from 150 to 200 $." It concluded with the request to "Donate Now: Telegram contact @AidForSYRIA1, Bitcoin [Wallet address] 1EutyMiNCEosn4xxAM9Avvir6fyWwUjdi, Plz Donate and spread the channel @AidForSYRIA."

On August 31, 2018, Al Ansaar forwarded a solicitation from the Telegram channel of the jihadi fundraising group Al Sadaqah, which also raises funds for fighters in Syria.[110] The solicitation stated: "There is a high possibility that the Syrian regime backed by Russia will be advancing into idilb province to retake the city very soon. We ask you to dig deep and assist the mujahideen in the defence of the last stronghold in Syria. Al Sadaqah would like to provide the mujahideen with equipment to stop any advance the regime

Al Ansaar - Aid For SYRIA
Forwarded from Al Sadaqah
There is a high possibility that the Syrian regime backed by Russia will be advancing into idlib province to retake the city very soon.

We ask you to dig deep and assist the mujahideen in the defence of the last stronghold in Syria.

Al Sadaqah would like to provide the mujahideen with equipment to stop any advance the regime make.

To make a secure and anoymous donation to the mujahideen in Idlib please contact us:

Telegram: @alsadaqahsyria
WhatsApp: +963 998 050 987

May Allah reward you and assist the mujahideen.   ⊙ 1   3:37 PM

133

make." It added: "To make a secure and anonymous donation to the mujaideen in Idlib please contact us: Telegram: @alsadaqahsyria WhatsApp: +963 998 050 987 May Allah reward you and assist the mujahideen."

Project Updates

On July 11, 2018, Al Ansaar posted a photo of a man holding an "Al Ansaar Aid for SYRIA" sign and some U.S. banknotes.

An August 7, 2018 post focused on two projects, "Ribaat/Houses Project" and "Houses of Brothers." Under "Houses of Brothers" it stated: "The 10 days of dhul hijjah are coming up where good deeds are multiplied... Our group has 10 houses in idlib. Almost every month between eating and spending and fuel is about $200 per house." Under "Ribat Project" it stated: "The Rabat fuel needs with eating with ammunition. Almost every point of $400 to $500 a month. We need split bags with drills in large quantities from $200 to $300. These are approximate numbers." It added: "To those who want to help, contacts are below, if you can't help financially, don't forget us in dua and please spread the message so that others can help. Donate Now: Telegram contact @AidForSYRIA1, Bitcoin [Wallet address] 1Euty-MiNCEosn4xxAM9Avvir6fyWwUjdi, Plz Donate and spread the channel @AidForSYRIA."

The Coming Storm: Terrorists Using Cryptocurrency

An August 9, 2018 post focused on "spending your wealth and how it will help you in front of Allah on that great day." It followed up by reiterating: "To those who want to help, contacts are below, if you can't help financially, don't forget us in dua and please spread the message so that others can help" and reiterated the information on how to donate.

An August 10, 2018 post titled "TRUE MEN WHO WILL LIVE AND DIE FOR THEIR RELIGION" stated: "Truly more important than wealth or anything else we can send in Al-laahs way is true MEN... How we long for each and every1 of our ummah to make the jihaad all over the world their jihaad and to protect and assist and guide it in the best of ways so that we may all get some share in the massive rewards that are available. Below is a video from yemen by a sheikh of the mujaahideen explaining to us what true men are. May Allaah make us of them." The post included a two-minute video of a man speaking in Arabic on "true men" according to Islam; the video was subtitled in English.

Al Ansaar - Aid For SYRIA
TRUE MEN WHO WILL LIVE AND DIE FOR THEIR RELIGION

Truly more important than wealth or anything else we can send in Allaahs way is true MEN.

MEN who will brave all the difficulties carrying this deen upon their shoulders until their last breath and drop of blood was given to Allaah and they meet him then rejoicing, overjoyed, delighted, at what he has given them from his bounties.

MEN who have sold this world in exchange for the aakhirah. And what a pitiful fool is the one who chooses this insignificant, temporary, transient worldly life over a jannah which is bigger than the heavens and the earth.

Aaah what a meeting it must be with ash shakoor when he saw you sacrifice tour family and wealth and efforts and time and blood for His sake. May He increase us in our love to meet Him and May He love to meet us.

How we long for each and every1 of our ummah to make the jihaad all over the world their jihaad and to protect and assist and guide it in the best of ways so that we may all get some share in the massive rewards that are available.

Below is a video from yemen by a sheikh of the mujaahideen explaining to us what true men are, May Allaah make us of them.

👁 71   12:10 AM

Al Ansaar - Aid For SYRIA

01:45

▶

👁 73   12:10 AM

The Coming Storm: Terrorists Using Cryptocurrency

An August 13, 2018 post included a hadith: "Zaid bin Khalid (May Allah be pleased with him) reported: The messenger of Allah said, 'He who equips a Ghazi (fighter) in the way of Allah is as if he has taken part in the fighting himself; and he would looks after the dependents (family) of a Ghazi in his absense, is as if he has taken part in the fighting himself.' [Al-Bukhari and Muslim] Don't miss out this golden opportunity dear brothers and sisters. Do your part from wherever you are." The post reiterated: "DONATIONS OR INFORMATION Telegram contact: @AidForSyria1."

On August 14, 2018, Al Ansaar wrote: "Right now northern syria is ardh al ribat and jihad [land of guard duty and jihad]. The Mujahideen are protecting borders here and the advancement of the *kuffar* [unbelievers] enemy is inevitable. So it is a defensive jihad which is obligated on every single able bodied Muslims by any means." The posts ended with "DONATIONS OR INFORMATION Telegram contact: @AidForSyria1."

On August 29, 2018, Al Ansaar wrote: "Alhamdullilah food was provided for brothers for some meals in a recent training camp... So know that your efforts are greatly appreciated in dunya [wordly life] and I ask Allaah to make all of you and us of those to whom it's is said to regarding Jannah [paradise]..." The post concluded with the standard information for contact or donations.

A lengthy September 1, 2018 post included a "letter from Swedish brothers to donor" that said in part: "Allah only gives His most beloved slaves the chance to be apart of this cause fighting with ones wealth to help mujahideen, we know so many brothers that never gave a penny because Allah did not make it easy for him, so it's a huge blessing. With that said we want to thank you from the bottom of our hearts and which that Allah will let you be from the people of jannatul firdaws [Paradise], the highest paradise being with the martyrs like the female *sahabiyyah* [female companions of the Prophet] that used to protect our beloved prophet Muhammad peace be upon him, Nusaybah Umm 'Imaarah, that lost her arm in one of the battle protecting him..." The post concluded with the standard information for contact or donations.

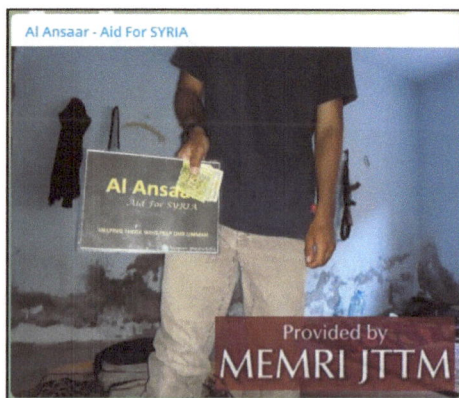

Also posted were photos of men holding up banknotes with the "Al Ansaar Aid For SYRIA" sign.

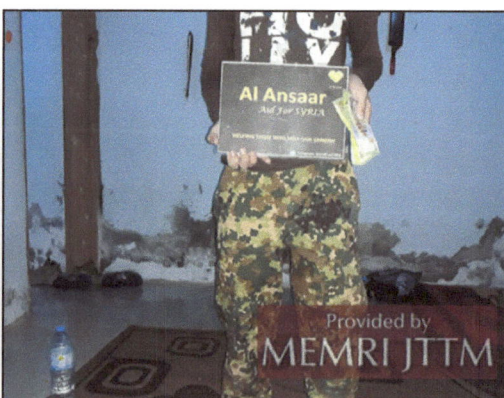

The Coming Storm: Terrorists Using Cryptocurrency

Jihadi "Independent Charity" Group Al-Ikhwa Continues To Solicit Funds On Telegram Via Bitcoin To Support Wounded Fighters, Wives Of 'Martyrs,' Orphans In Syria: 'Earn Your Tickets To Jannah [Paradise] By Donating'

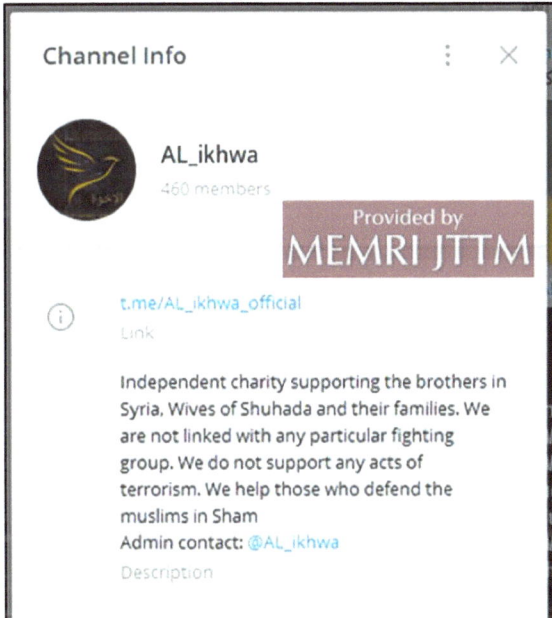

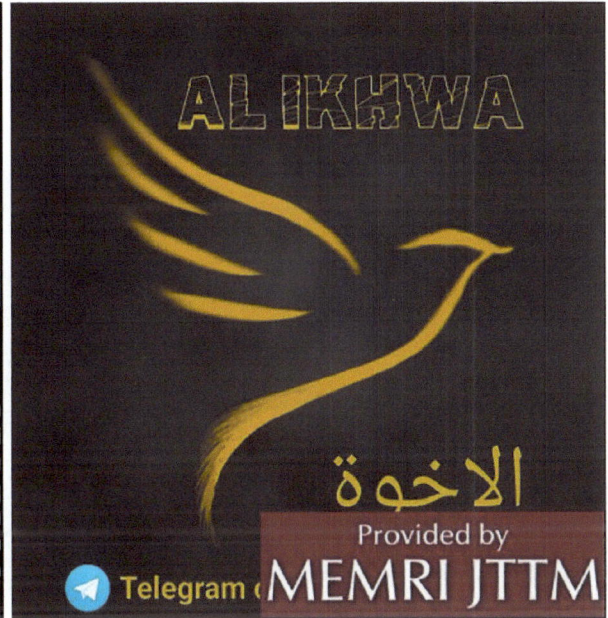

Introduction

"Disclaimer" on the Al-Ikhwa channel (June 6, 2018).

The Al-Ikhwa ("The Brothers") Telegram channel, launched June 5, 2018 and currently with 460 members,[111] describes itself in its Channel Info as: "Independent charity supporting the brothers in Syria. Wives of Shuhada [Martyrs] and their families. We are not linked with any particular fighting group. We do not support any acts of terrorism. We help those who defend the Muslims in sham [i.e., Syria]."[112] The channel's admin is @AL_ikhwa.

The following are examples of posts on the AL_ikhwa Telegram channel from June through November, 2018.

139

General Appeals For Funds

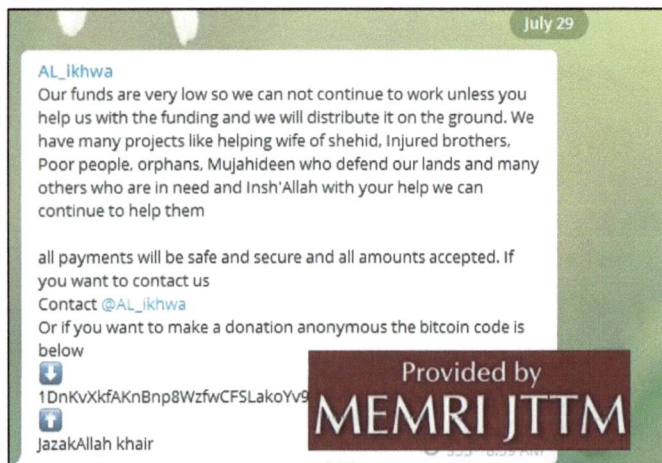

Appeal for funds (July 29, 2018).

On June 9, 2018, the group posted an appeal for funds: "We are working on some good projects... And we cannot do it without your support so please donate and help your brothers and sisters. Also, we are sorting out some better payment methods for those who want to stay anonymous. Forgive us and just bear with us to set this up in the near future..."

On July 4, the group began accepting donations via bitcoin. It posted its bitcoin address and wrote: "We have a bitcoin now for anyone who wants to donate to our causes. Funding is limited so we cannot continue any of these projects without your support... All payments will be safe and secure, and all amounts accepted..." As of this writing, the bitcoin address shows zero transactions.[113]

On July 29, the group again noted its need for donations in order to continue its operations, and asked that they be in bitcoin. It wrote: "Our funds are very low, so we cannot continue to work unless you help us with the funding and we will distribute it on the ground. We have many projects like helping wife of shehid [martyr], injured brothers, poor people, orphans, Mujahideen who defend our lands and many others who are in need..."

Al-Ikhwa's appeals for donations are often placed in an Islamic context. On June 6, the group published two posters, reiterating the divine reward and punishment for those who do and do not donate.

The Coming Storm: Terrorists Using Cryptocurrency

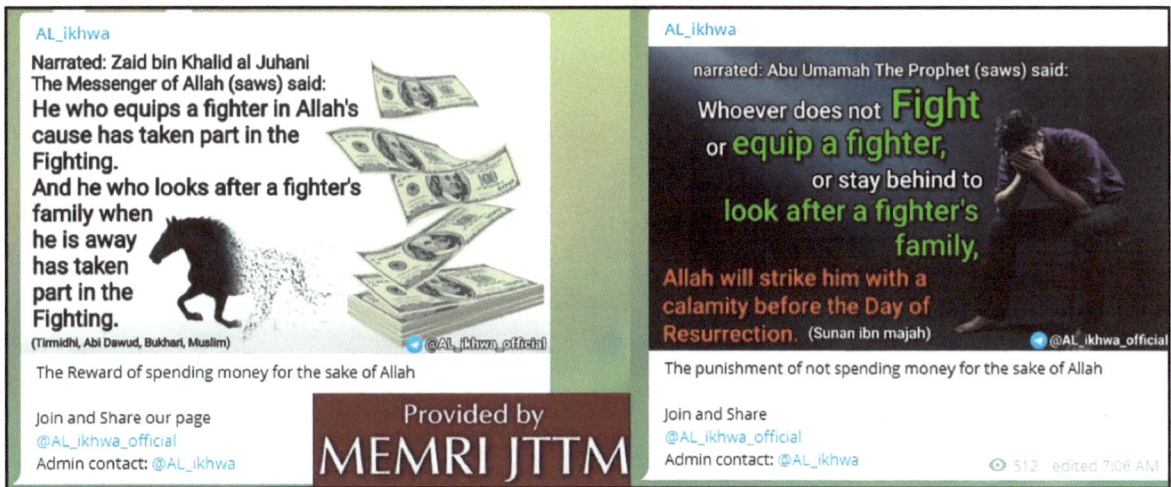

Posters citing hadiths promising divine rewards for those who donate and punishment for those who do not (June 6, 2018).

On September 9, the channel posted a photo of AL_ikhwa's empty Bitcoin Personal Wallet, which states: "It's a ghost town in here." It is captioned: "Even our bitcoin is making fun of our low funding But with your help we can continue these charity projects Insh.'Allah So Please DONATE NOW."

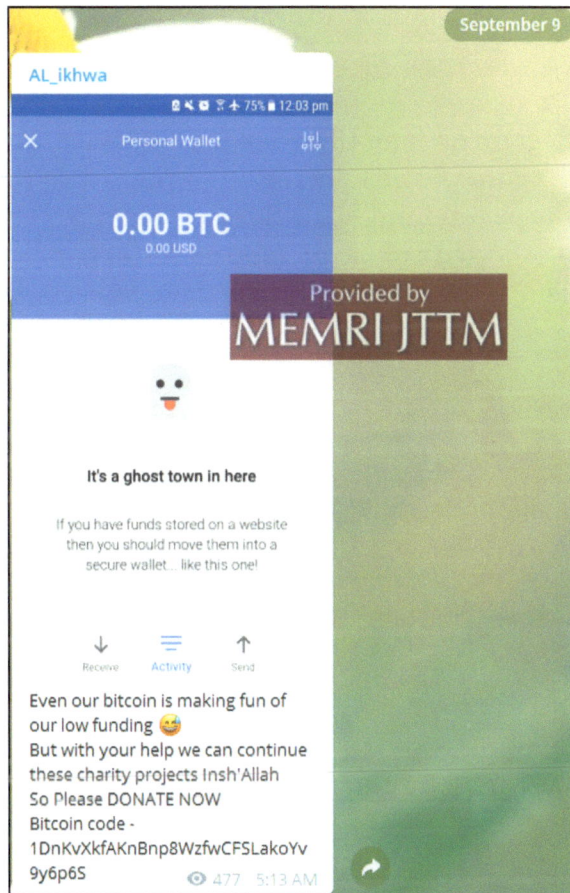

Stalinsky ▪ M E M R I

According to its Telegram channel, the donations received have been used to provide meals for "brothers who have been doing ribat (guarding in frontline)" and for supporting wounded "brothers." This support includes providing a prosthetic leg for one wounded in an explosion, and for another who is now a paraplegic following a gunshot wound. The photos of the goods provided feature AL_ikhwa signs.

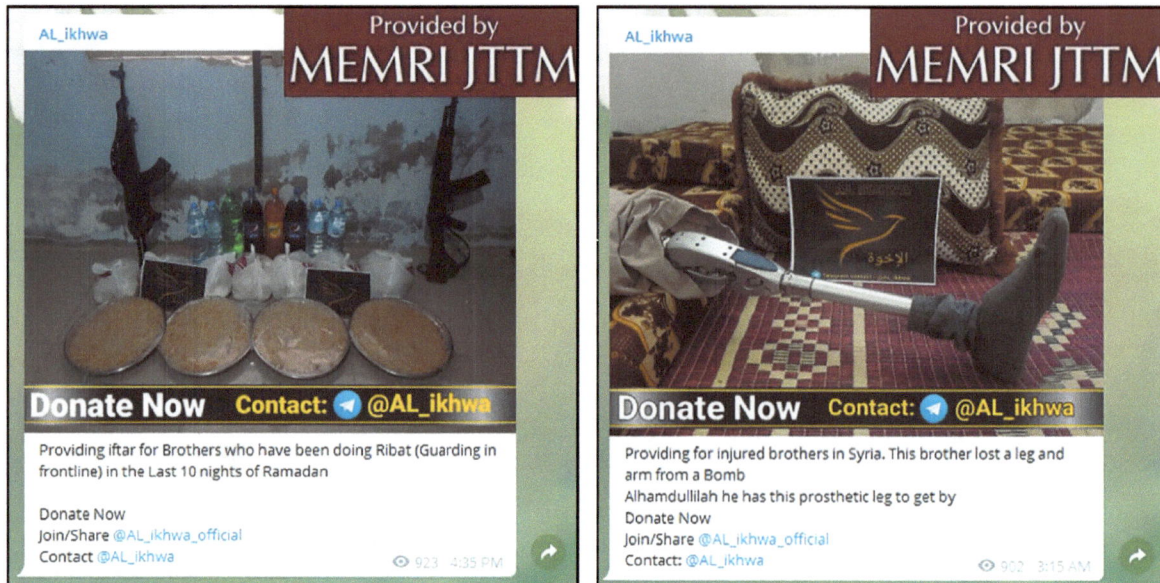

Providing food for fighters on frontlines (June 10, 2018); providing a prosthetic leg (June 11, 2018).

AL_ikhwa posted on Telegram on June 18 asking for donations, saying: "Remember whatever you spend in Allahs cause will be fully repaid to you... with these donations...you are also helping yourself with rewards from Allah. So earn your tickets to Jannah by donating to our causes..." On June 23, it posted a photo of "food packs for families in need" accompanied by an AL_ikhwa sign, adding: "We cannot continue these projects without your donations."

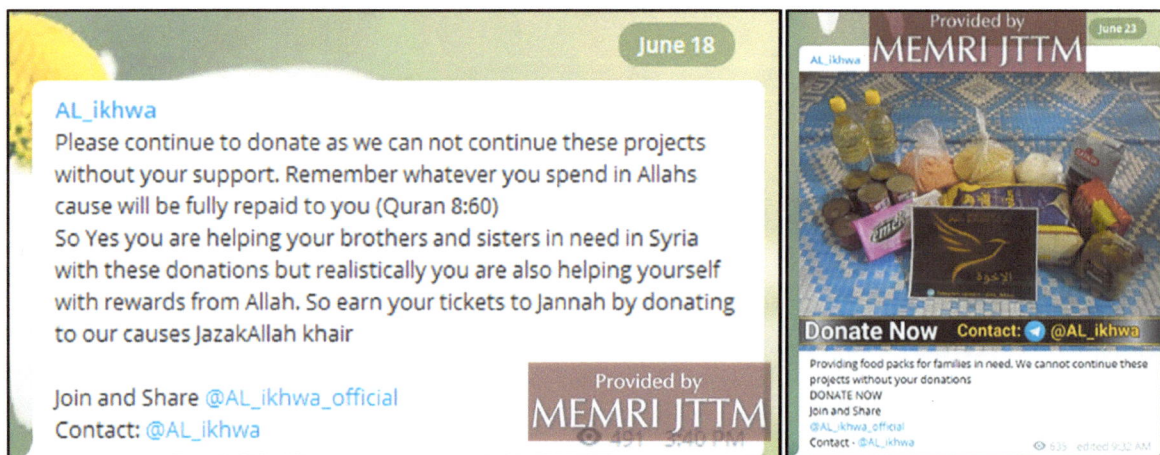

The Coming Storm: Terrorists Using Cryptocurrency

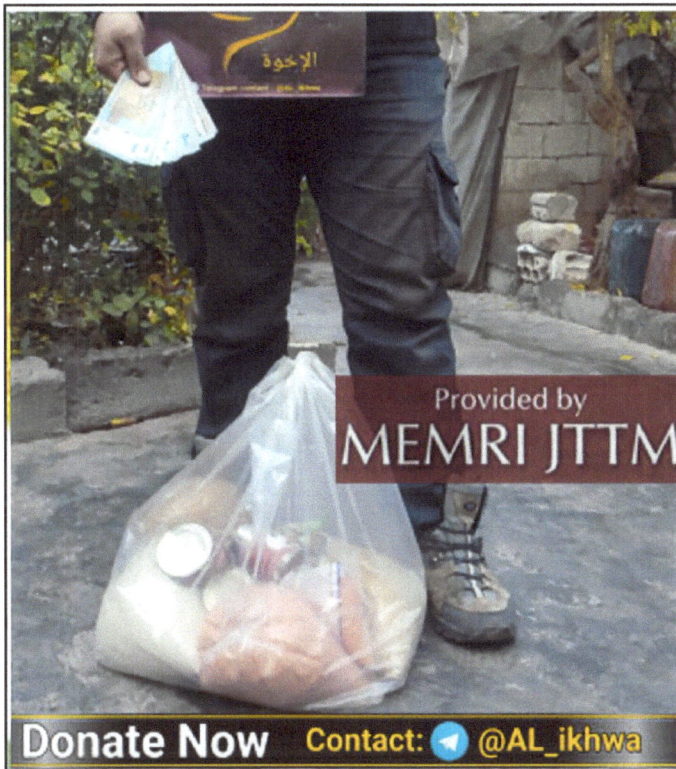

This brother –
like all muhajireen left their countries, families,
wealth, and risked their lives to come here to
defend the Muslims of Sham.
Everyone who comes here expects to fight and
help the Ummah and their life will be funded by
the group they work with.
But months pass with no money and the bills
keep coming.
We heard this brother had no money and he got
a job picking olives just to pay the bills. Having a
job is a good thing.
We all need to work and even the Sahabas
balanced their fighting life with their work life. The
problem is the Jobs pay little here so this brother
found no time for his Jihad life.
Which is the whole reason he made Hijra in the
first place. Alhamdullilah we were able to help this
brother by giving him some money and a food
pack

There is many similar cases like this here
SO DONATE NOW
and dont forget your brothers and sisters in Syria
Join and Share
@AL_ikhwa_official
Contact - @AL_ikhwa
Anonymous donation available Bitcoin -
1A4kjy59YcYJNThyEX9nGgj7615qnXVgea

Provided by
MEMRI JTTM

Donate Now   Contact: @AL_ikhwa

On November 19, AL_ikhwa posted a photo of a man holding currency and an an AL_ikhwa sign with a package of what appears to be donated materials at his feet. The post states: "This brother – like all muhajireen left their countries, families, wealth, and risked their lives to come her to defend the Muslims of Sham. Everyone who comes here expects to fight and help the Ummah and their life will be funded by the group they work with... The problem is the jobs pay little here so this brother found no time for his Jihad life."

On September 26, the Telegram channel posted a photo of a woman with a package of food holding an AL_ikhwa sign. The caption says: "Providing food packs for families in need We cannot continue these projects without your support."

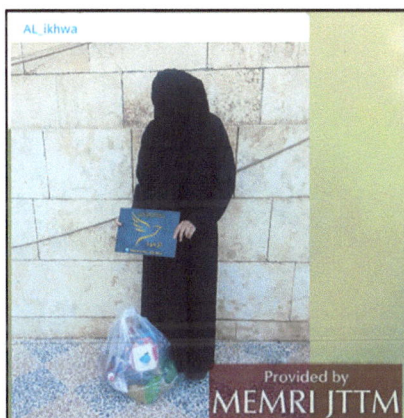

Donate Now   Contact: @AL_ikhwa

Providing food packs for families in need
We cannot continue these projects without your
support
@AL_ikhwa_official
Contact - @AL_ikhwa
DONATE NOW
Bitcoin donation
1DnKvXkfAKnBnp8WzfwCFSLakoYv9y6p6S

👁 327   12:37 PM

Provided by
MEMRI JTTM

September 29

AL_ikhwa posted a photo of a child holding up a sign with their logo on August 14. The caption reads: "Providing food packs for families in need / This little girl was too shy for a photo so she wore her mothers Niqab."

The Coming Storm: Terrorists Using Cryptocurrency

On June 25, Al-Ikhwa posted a picture of a "wife of shehid" and children, saying she was a *muhajira* ("immigrant") whose husband had been killed several months ago, and that she had five children. Noting that there were many other similar cases, it again requested donations: "She is a *muhajira* and her husband got killed a few months ago and she is left alone with her 5 children and no support. There is many similar cases like this. This is a duty upon the *ummah* to help our brothers and sisters like this. Her husband gave his life for the sake of Allah and to help defend the Muslims from the criminal regime and its allies. Now it's our turn to return the favor. Donate now. We cannot continue these projects without your support."

On June 9, the channel posted a request for donations to help "your brothers and sisters." It adds: "We are sorting out some better payment methods for those who want to stay anonymous... but Insh'Allah we will do our best to sort out this as soon as possible JazakAllah khair for now please keep joining and sharing out page."

Donating During Eid

AL_ikhwa

Eid Mubarak to all who follow our channel and please share to others our greeting.

May Allah accept from all of us our good deeds and forgive us our sins. Ameen

May you all enjoy your time with your families this Eid.

But do not forget those other families who are in need. Or the homeless poor muslims. Or the orphans who lost their families. Or wives of shehid who lost their husbands. Dont forget them in your Dua or If you want to help them by sending money the contact is below. Or if you want to stay anonymous by paying bitcoin is also available.

We have very low funds here in Syria and we can only work and continue these projects if you help us with the funding. You fund it and we distribute it

Please join and share our page
@AL_ikhwa_official
DONATE NOW contact us on
@AL_ikhwa
Bitcoin donation 1DnKvXkfAKnBnp8WzfwCFSLakoYv9y6p6S
JazakAllah khair.   👁 656   3:33 AM

Provided by MEMRI JTTM

On August 21, the channel posted: "May you all enjoy your time with your families this Eid. But do not forget those other families who are in need. Or the homeless poor muslims. Or the orphans who lost their families. Or wives of shehid who lost their husbands. Don't forget them in your Dua or if you want to help them by sending money the contact is below. Or if you want to stay anonymous by paying Bitcoin is also available." The post also provided AL_ikhwa's bitcoin address.

On August 21, AL_ikhwa posted a photo of bags filled with treats with the AL_ikhwa label on them. The photo is captioned: "Giving the children Eid Snack bags. Chips, chocolates and lollies We cannot continue these projects without your support. DONATE NOW." Below is a photo of four little boys holding snack bags; two of them also hold small automatic weapons.

The Coming Storm: Terrorists Using Cryptocurrency

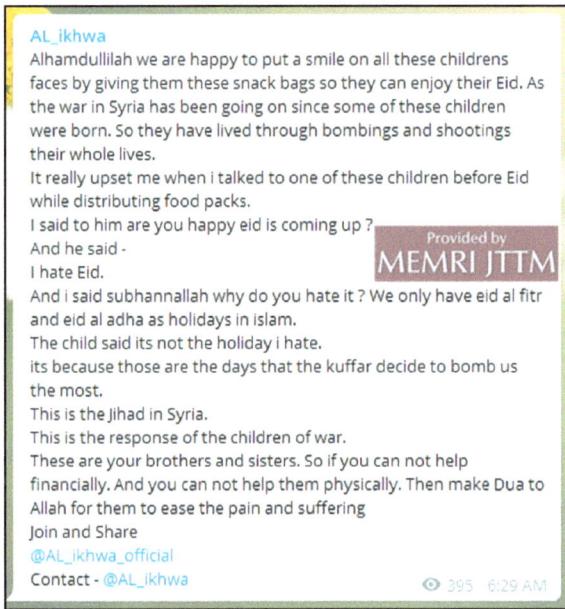

AL_ikhwa
Alhamdullilah we are happy to put a smile on all these childrens faces by giving them these snack bags so they can enjoy their Eid. As the war in Syria has been going on since some of these children were born. So they have lived through bombings and shootings their whole lives.
It really upset me when i talked to one of these children before Eid while distributing food packs.
I said to him are you happy eid is coming up ?
And he said -
I hate Eid.
And i said subhannallah why do you hate it ? We only have eid al fitr and eid al adha as holidays in islam.
The child said its not the holiday i hate.
its because those are the days that the kuffar decide to bomb us the most.
This is the Jihad in Syria.
This is the response of the children of war.
These are your brothers and sisters. So if you can not help financially. And you can not help them physically. Then make Dua to Allah for them to ease the pain and suffering
Join and Share
@AL_ikhwa_official
Contact - @AL_ikhwa
395 6:29 AM

On August 21, the channel described an interaction with a Syrian boy who claims to "hate Eid" because "those are the days that the kuffar decide to bomb us the most." The post reads: "This is the Jihad in Syria. This is the response of the children of war. These are your brothers and sisters. So if you can not help financially. And you can not help with physically. Then make Dua..."

Monetary Donations

On September 2, AL_ikhwa posted a hadith stating "There will be people from my nation who will protect the borders. They will fulfill their duty but they will not be given what is due to them." This, it adds, "fits our time" as "a lot of mujahideen go to ribat, battles, trainings, meetings, etc and are expected to be on call and ready. But the groups neglect them and don't pay them their monthly wage sometimes for many months." The post concludes by saying: "Remember the hadith the one who funds a fighter has fought...there is no better reward in islam than Jihad fisabilillah. So you get that reward insh'Allah by funding a brother."

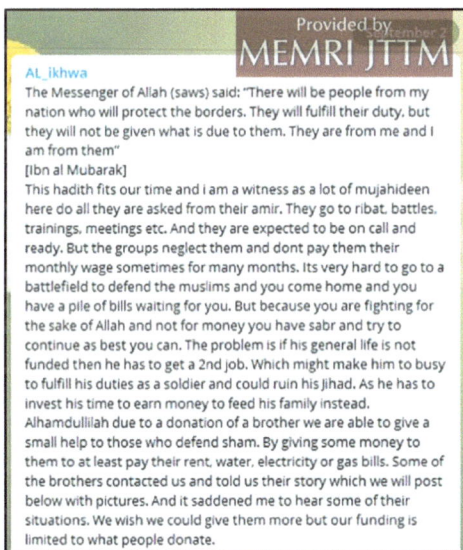

AL_ikhwa
The Messenger of Allah (saws) said: "There will be people from my nation who will protect the borders. They will fulfill their duty, but they will not be given what is due to them. They are from me and I am from them"
[Ibn al Mubarak]
This hadith fits our time and i am a witness as a lot of mujahideen here do all they are asked from their amir. They go to ribat, battles, trainings, meetings etc. And they are expected to be on call and ready. But the groups neglect them and dont pay them their monthly wage sometimes for many months. Its very hard to go to a battlefield to defend the muslims and you come home and you have a pile of bills waiting for you. But because you are fighting for the sake of Allah and not for money you have sabr and try to continue as best you can. The problem is if his general life is not funded then he has to get a 2nd job. Which might make him to busy to fulfill his duties as a soldier and could ruin his Jihad. As he has to invest his time to earn money to feed his family instead.
Alhamdullilah due to a donation of a brother we are able to give a small help to those who defend sham. By giving some money to them to at least pay their rent, water, electricity or gas bills. Some of the brothers contacted us and told us their story which we will post below with pictures. And it saddened me to hear some of their situations. We wish we could give them more but our funding is limited to what people donate.

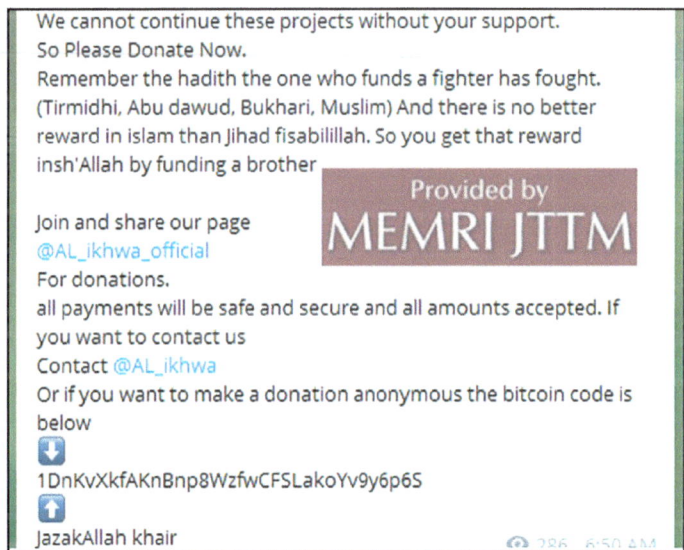

We cannot continue these projects without your support.
So Please Donate Now.
Remember the hadith the one who funds a fighter has fought. (Tirmidhi, Abu dawud, Bukhari, Muslim) And there is no better reward in islam than Jihad fisabilillah. So you get that reward insh'Allah by funding a brother
Join and share our page
@AL_ikhwa_official
For donations.
all payments will be safe and secure and all amounts accepted. If you want to contact us
Contact @AL_ikhwa
Or if you want to make a donation anonymous the bitcoin code is below
1DnKvXkfAKnBnp8WzfwCFSLakoYv9y6p6S
JazakAllah khair
286 6:50 AM

A September 2 post includes a photo of currency with an AL_ikhwa sign, and adds: "Helping the Mujahideen with Rent, Gas, Electricity and general living costs."

AL_ikhwa
The Messenger of Allah (saws) said: "There will be people from m...

Provided by
MEMRI JTTM

AL IKHWA
الإخوة
Telegram contact - @AL_ikhwa

Donate Now   Contact: 📲 **@AL_ikhwa**

Helping the Mujahideen with Rent, Gas, Electricity and general living costs
Join and Share
@AL_ikhwa_official
Contact - @AL_ikhwa

👁 238   6:52 AM

In another September 2 post, AL_ikhwa featured a man holding currency, a rifle, and an AL_ikhwa sign, adding: "Newly married mujahid we were able to give some money to help out with his rent."

The Coming Storm: Terrorists Using Cryptocurrency

AL_ikhwa

AL_ikhwa

The Messenger of Allah (saws) said: "There will ...

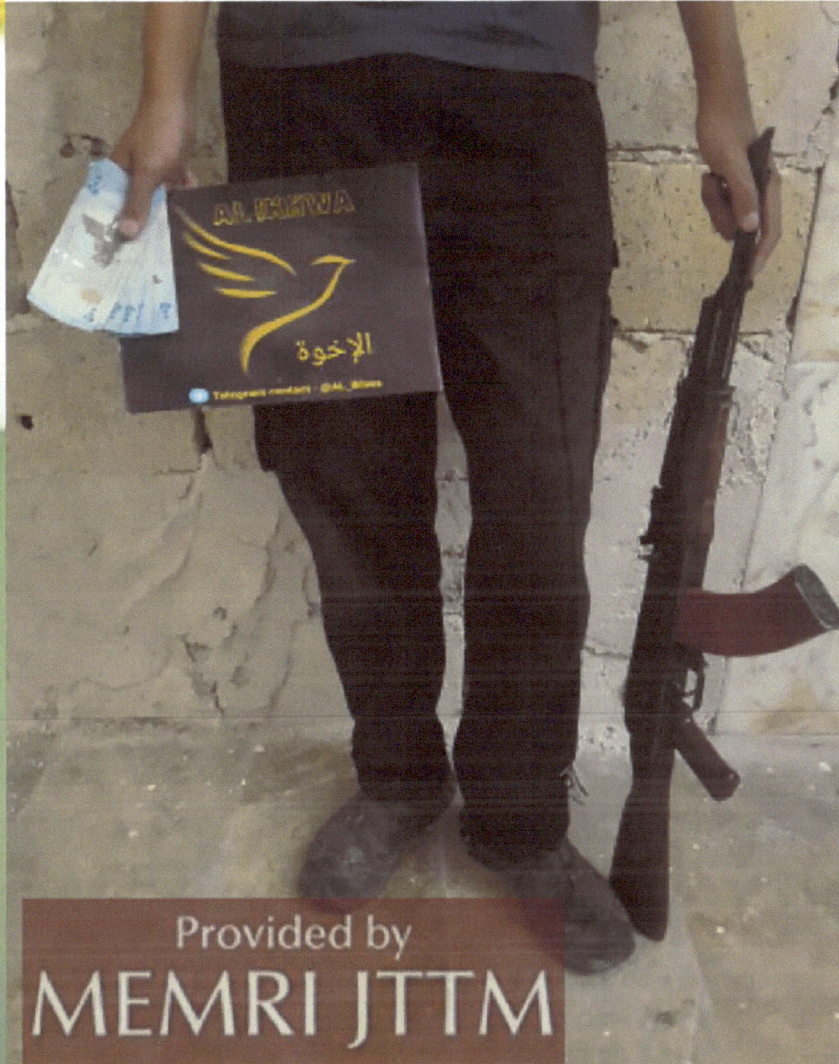

Provided by

MEMRI JTTM

Donate Now Contact: @AL_ikhwa

Newly married mujahid we were able to give
some money to help out with his rent.
Join and Share
@AL_ikhwa_official
Contact - @AL_ikhwa

👁 360   6:59 AM

Also on September 2, AL_ikhwa posted a story about a mujahid who "had no money to pay for... petrol with money" so was in debt but who was given money by AL_ikhwa to "ease his affairs" as well as a "food pack so he can feed his family." It adds that "'the only thing that stops the shehid from entering Jannah is Debt,' so even though he gave his life to allah, "he's barred from entering Jannah until his debt is cleared."

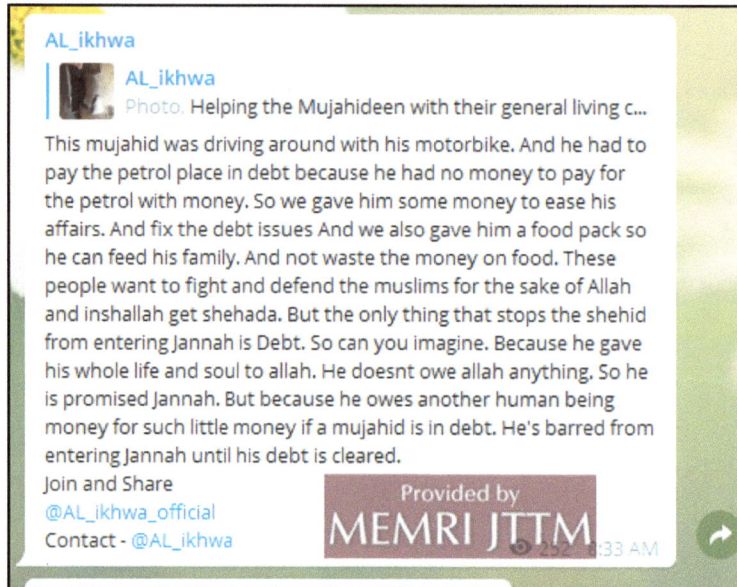

AL_ikhwa

AL_ikhwa
Photo. Helping the Mujahideen with their general living c...

This mujahid was driving around with his motorbike. And he had to pay the petrol place in debt because he had no money to pay for the petrol with money. So we gave him some money to ease his affairs. And fix the debt issues And we also gave him a food pack so he can feed his family. And not waste the money on food. These people want to fight and defend the muslims for the sake of Allah and inshallah get shehada. But the only thing that stops the shehid from entering Jannah is Debt. So can you imagine. Because he gave his whole life and soul to allah. He doesnt owe allah anything. So he is promised Jannah. But because he owes another human being money for such little money if a mujahid is in debt. He's barred from entering Jannah until his debt is cleared.
Join and Share
@AL_ikhwa_official
Contact - @AL_ikhwa

Provided by
MEMRI JTTM

On June 25, AL_ikhwa wrote about its assistance to "wives of the shehid" and explains how it had "collected money" for one with "5 children one which a 1 month old baby" whose "husband got killed a few months ago." It adds: "This is a duty upon the ummah to help our brothers and sisters like this. Her husband gave his life for the sake of Allah and to help defend the muslims from the criminal regime and its allies."

AL_ikhwa

AL_ikhwa
Photo. Providing for the wife of shehid. We cannot contin...

Helping the wives of the shehid
This women has 5 children one which is a 1month old baby. She was in need of baby milk and nappies so we collected money for her to buy what is needed for her children. She is a muhajira and her husband got killed a few months ago and she is left alone with her 5 children and no support. There is many similar cases like this. This is a duty upon the ummah to help our brothers and sisters like this. Her husband gave his life for the sake of Allah and to help defend the muslims from the criminal regime and its allies. now its our turn to return the favour. Donate now. We cannot continue these projects without your support
Join and Share
@AL_ikhwa_official
Contact: @AL_ikhwa

Provided by
MEMRI JTTM

1927 7:53 AM

July 2

The Coming Storm: Terrorists Using Cryptocurrency

A photo of what appears to be utility bills and an AL_ikhwa sign was posted on September 9, with the caption:: "Helping the Mujahideen with Rent, Gas, Electricity and general living costs."

On October 3, AL_ikhwa posted a description of its distribution of collection boxes to shops in Idlib. Noting that "here the people don't have much compared to outside," the channel expressed the hope that people will take this opportunity "to join in and be listed as those who supported the Jihad."

The Coming Storm: Terrorists Using Cryptocurrency

On October 8, the channel posted a photo of a woman holding an AL_ikhwa envelope with currency in it, captioned: "Providing some money for families in need We cannot continue these projects without your support." The post ends with the Contact - @AL_ikhwa and the Bitcoin address.

October 8

AL_ikhwa

Provided by
MEMRI JTTM

Donate Now   Contact: @AL_ikhwa

Providing some money for families in need
We cannot continue these projects without your support
@AL_ikhwa_official
Contact - @AL_ikhwa
DONATE NOW
Bitcoin donation 1DnKvXkfAKnBnp8WzfwCFSLakoYv9y6p6S

👁 388   8:00 AM

On September 15, the channel forwarded a graphic from "Reminders From Syria" stating: "Charity does not in any way decrease your wealth." It adds: "People don't give charity as they fear they lose money But do we trust The Prophet saws? Its actually the opposite. When you give in charity Allah gives you more back."

The Coming Storm: Terrorists Using Cryptocurrency

Indonesian Jihadi Fundraising Group Abu Ahmed Foundation Gives Bank Account Numbers, Works With Other Jihadi Organizations And Uses Cryptocurrencies

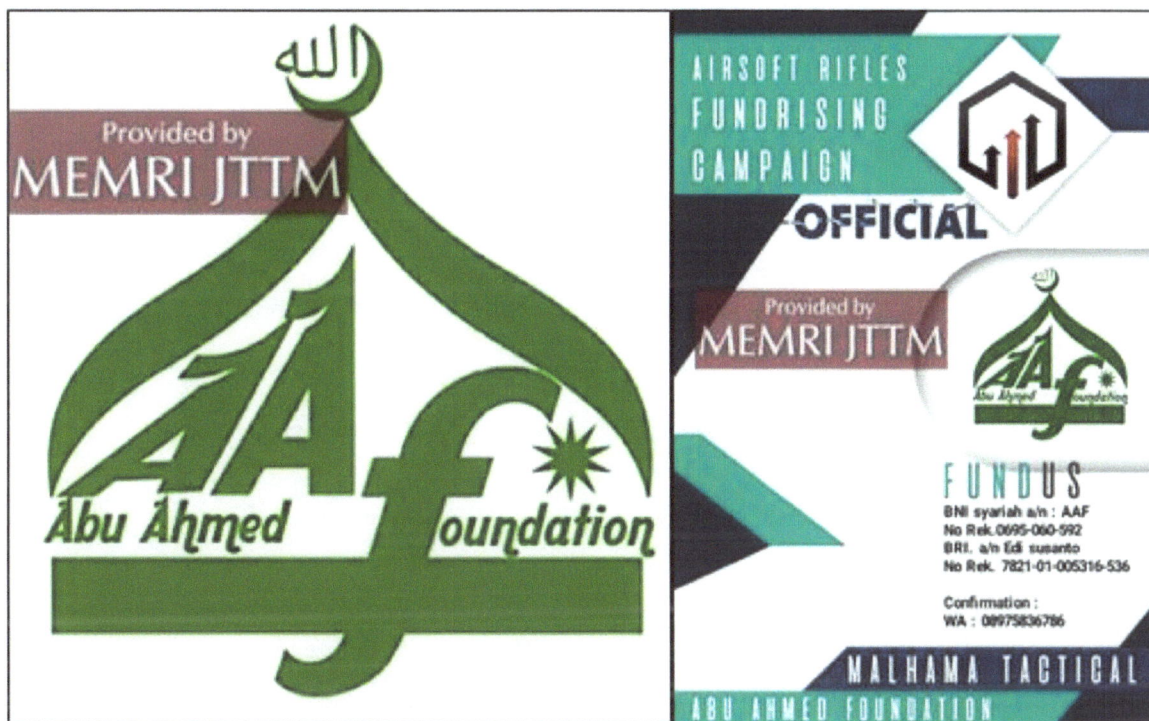

On October 21, 2018, the Abu Ahmed Foundation (AAF), an Indonesian fundraising group that raises money to buy food, weapons, clothing, and other materials for jihadis, opened a new Telegram channel. AAF frequently posts, primarily in Indonesian, on its Facebook, Twitter, WhatsApp, and Instagram accounts. The group's Twitter and Facebook pages say it is located in Syria, while the group's other two social media accounts do not specify a location. Many of the videos and photos that the group posts appear to have been taken in Indonesia, while others appear to have been taken in Syria.

According to its social media posts, AAF works with Syria-based jihadi training group Malhama Tactical on multiple projects including a campaign to raise funds for airsoft rifles. AAF gives a WhatsApp contact number, 08975836786, and bank accounts with Indonesian banks BNI Syariah and Bank Rakyat Indonesia. The group also posts pictures of protesters carrying Hay'at Tahrir Al-Sham (HTS) flags.

In April 2018, MEMRI JTTM published a report about AAF discussing how AAF collaborates with other jihadi fundraising groups, such as Al-Sadaqah Foundation and The Merciful Hands. The report also covers how the group had previously listed its WhatsApp number as +963 935 740 232, and how it had encouraged supporters to donate using the Bitcoin, Monero, Dash, and Verge cryptocurrencies.

This section will review AAF's social media presence as well as how the group raises and spends funds.

Key Findings By JTTM Team

The MEMRI JTTM team has found numerous social media accounts, as well as contact and payment information connected to Abu Ahmed Foundation. AAF is present on Instagram (@abu_ahmed_foundation_), Twitter (@AbuFoundation), Facebook (Abu Ahmed Foundation), Telegram (NEW AAF MEDIA CENTRE), and WhatsApp (08975836786). The group has posted its WhatsApp phone number on its Telegram channel and Instagram profile. The same number is also listed on Facebook as the group's contact number. On Telegram and Instagram, the group has also shared banking and payment information for donations, naming two Indonesian banks: BNI Syariah (a/n AAF, No rek.0695-060-592) and Bank Rakyat Indonesia (BRI a/n Edi Susanto, No rek.7821-01-005316-536).

AAF ACCOUNT INFORMATION

Provided by
MEMRI JTTM

- 08975836786
- 963 935 740 232
- Bank Rakyat Indonesia (BRI a/n Edi Susanto, No rek.7821-01-005316-536)
- BNI Syariah (a/n AAF, No rek.0695-060-592)
- @AbuFoundation
- Abu Ahmed Foundation
- @abu_ahmed_foundation_
- NEW AAF MEDIA CENTRE

The Coming Storm: Terrorists Using Cryptocurrency

AAF Telegram Channel

On October 21, 2018, Abu Ahmed Foundation created a new Telegram channel called "NEW AAF MEDIA CENTRE." According to the first message posted on the channel, the group's previous account had been suspended.

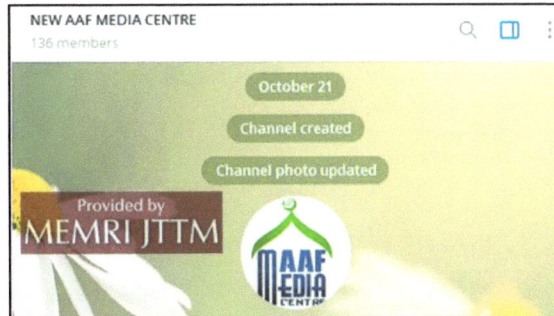

On October 23, two days after it was created, the channel had 198 members, 69 photos, 16 videos, and three shared links.

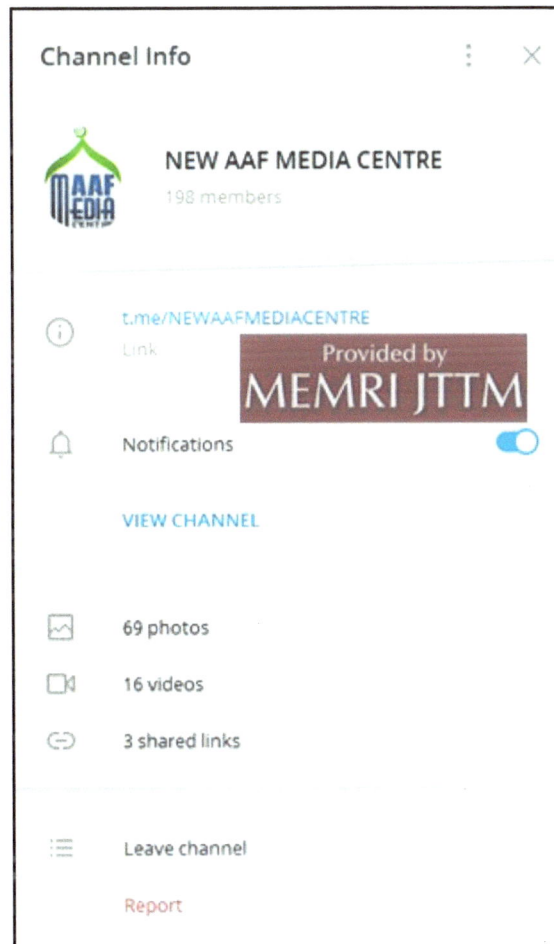

The first post in the channel read: "Bismillah, our new Channel Alhamdulilah after suspension, AAF wish to thank all for supporting us over the last two years and now ask for your help again to help rebuild our chanel again." The group encouraged readers to donate to frontline mujahideen: "If you cant do jihad in person, then equip the mujahidin and their familys*." The post continues to explain how the program is usually divided into three parts: "*mujahid and *ribath* donation/*infaq* *family of *syuhada* [martyrs] and active mujahid and orphans* then *education and dakwa [sic *da'wa*, preaching].*"

NEW AAF MEDIA CENTRE
Aisyah:

Bismillah, our new Channel Alhamdulillah after suspension, AAF wish to thank all for suporting us over the last two years and now ask for your help again to help build our chanel again, AAF your bridge to the frontline mujahedeen inshaaallah, carrying out Allah ta'ala law, *if you cant do jihad in person, then equip the mujahidin and their familys* our program is usually devided into three according to the priority in sheria islam, *mujahid and ribath donation/infaq* *family of syuhada and active mujahid and orphans* then *education and dakwa* inshaaAllah, jazakallah kahiran... Help us to help the wheels of jihad turn inshaaAllah... assalamualaikum wr wb (Ummu jibril Founder AAF)

👁 96   12:54 AM

Provided by MEMRI JTTM

On October 21, the group posted a second message introducing what appeared to be a joint project with Malhama Tactical. The message read: "Bismallah... help AAF to help them set up Airsoft training camp which will be open for training mujahid in elite skills inshaaallah in safe environment with latest equipment...If you can't do jihad right now? Do the next best things with the same reward inshaaallah *jihad with wealth*."

MALHAMA TACTICAL   ABU AHMED FOUNDATION

BNI syariah a/n : AAF
No Rek.0695-060-592
BRI. a/n Edi susanto
No Rek. 7821-01-005316-536

Confirmation :
WA : 08975836786

Provided by MEMRI JTTM

AIRSOFT RIFLES   FUNDRISING CAMPAIGN

The Coming Storm: Terrorists Using Cryptocurrency

On April 9, 2018, an AAF Telegram channel that is no longer available shared a graphic that indicates a cooperation with jihadi fundraising group Al-Sadaqah and encourages supporters to donate using the Bitcoin, Monero, Dash, and Verge cryptocurrencies. The graphic also gives the WhatsApp number +963 935 740 232, the AAF Facebook page and unavailable Telegram channel, the Al-Sadaqah Telegram channel @alsadaqahsyria, and the Al-Sadaqah address on the encrypted email service ProtonMail alsadaqah@protonmail.com.

On October 21, AAF posted on its active Telegram channel a photo of a t-shirt bearing the group's emblem, something in a package bearing Indonesian writing, and five $100 bills.

NEW AAF MEDIA CENTRE

Provided by
MEMRI JTTM

TERIMA KASIH
SEPATUNYA
SAUDARA
INDONESIA

#AAF

The Coming Storm: Terrorists Using Cryptocurrency

Photos of many uniformed fighters were posted on October 21. The fighters' faces are blurred.

On October 21, the group posted photos of fighters holding up empty packages. The labels on the packages read: "AAF Media." The packages may have contained the uniforms that the fighters are wearing in the photo.

The same day, the group posted a photo of a man wearing an AAF shirt and holding two pairs of boots.

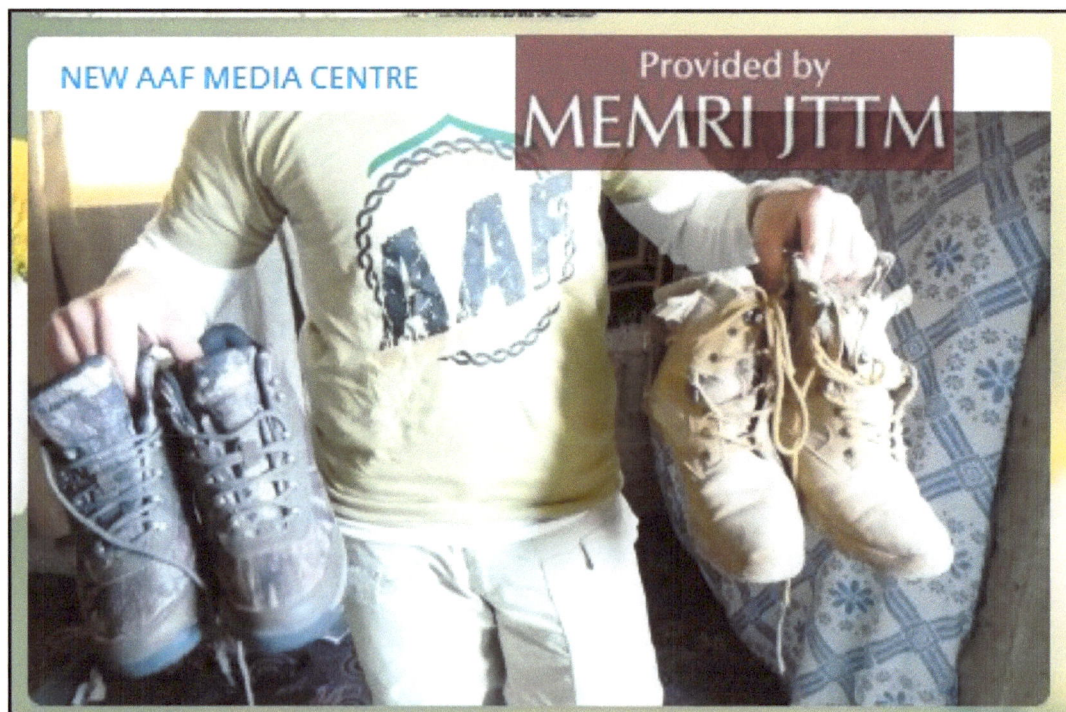

The Coming Storm: Terrorists Using Cryptocurrency

Also, on October 21, the channel posted photos of a line of boots. One of the pairs of boots has a tag bearing the price $49.98.

The same day, AAF posted a message in Indonesian giving the group's WhatsApp number, 08975836786, and discussing Malhama Tactical. The message also gives what appear to be bank account numbers for two Indonesian banks: BNI Syariah, a/n: AAF, 0695-060-592 and Bank Rakyat Indonesia, a/n: Edi Susanto, and 7821-01-005316-536.

NEW AAF MEDIA CENTRE

NEW AAF MEDIA CENTRE
Photo

Bismillah.

sebuah kehormatan besar bgi kami tim AAF karna mendapatkan penawaran dari akhi abu salman al balaros amir dari tim" Malhama Tactical " beliau ingin membuat suatu kerja sama dengan AAF untuk memenuhi siapan keperluan airsoft gun yang mana ini digunakan sebagai keperluan latihan tim elite. oleh sebab itu kami tim aaf membuka infaq bagi keperluan ini, insyaallah akan menjadi amal jariah bagi kaum muslimin sebagai persiapan mujahidin. bagi para kaum muslimin yang ingin berinfak dan berniat untuk membantu saudara kita silahkan melalui:

BNI syariah a/n : AAF
No rek.0695-060-592
BRI. a/n Edi susanto
No rek. 7821-01-005316-536

confirmasi : WA : 08975836786

jazakallah khoir 😊

#AAF
#ABUAHMEDFOUNDATION
#MALHAMATACTICAL

👁 157   uvuvava vev..., 6:42 AM

The group posted a photo of a person holding up a poster showing a flag bearing the *shahada* (i.e., "There is no god but Allah and Muhammad is his messenger") along with a sign that read: "Al Ansaar," which is the name of another jihadi fundraising group that uses Bitcoin, Telegram, and WhatsApp to collect money to supply the mujahideen with equipment such as weapons and drones, as well as food and clothing. There is a bag of items sitting underneath the flag.

The Coming Storm: Terrorists Using Cryptocurrency

On October 21, the group posted to tell members that "all 12 food packs have been delivered" and that the families of the mujahideen struggle with "basic provisions."

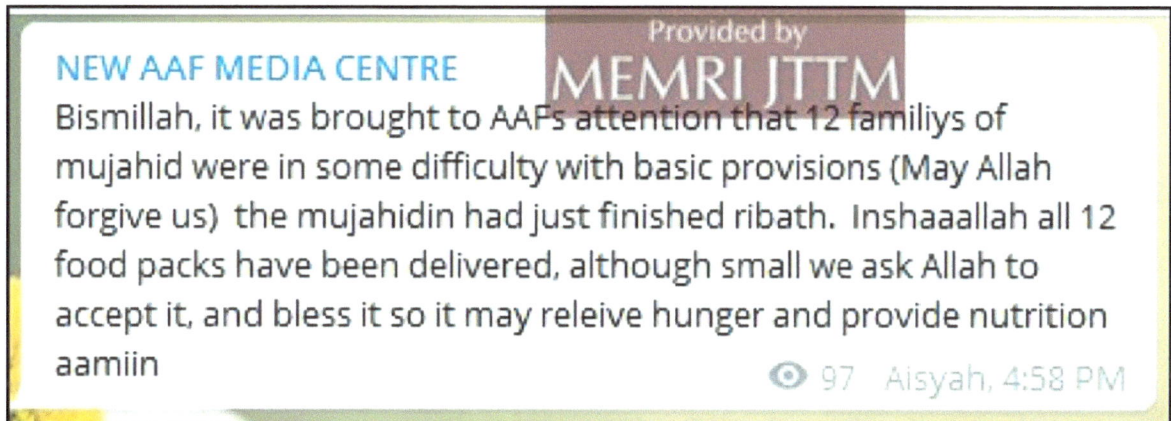

NEW AAF MEDIA CENTRE

Bismillah, it was brought to AAFs attention that 12 familiys of mujahid were in some difficulty with basic provisions (May Allah forgive us) the mujahidin had just finished ribath. Inshaaallah all 12 food packs have been delivered, although small we ask Allah to accept it, and bless it so it may releive hunger and provide nutrition aamiin

👁 97   Aisyah, 4:58 PM

On April 9, the group wrote on its Facebook page: "Open Order For Donation" and posted photos of shirts that read: "We support AAF."

Another post promotes the AAF shirts and gives a WhatsApp contact number.

On March 8, AAF posted a photo on Instagram of a sign that read: "AAF – MBP FOR GH-OUTA" and a few $100 bills. MBP stands for "Muslimah Bima Peduli."

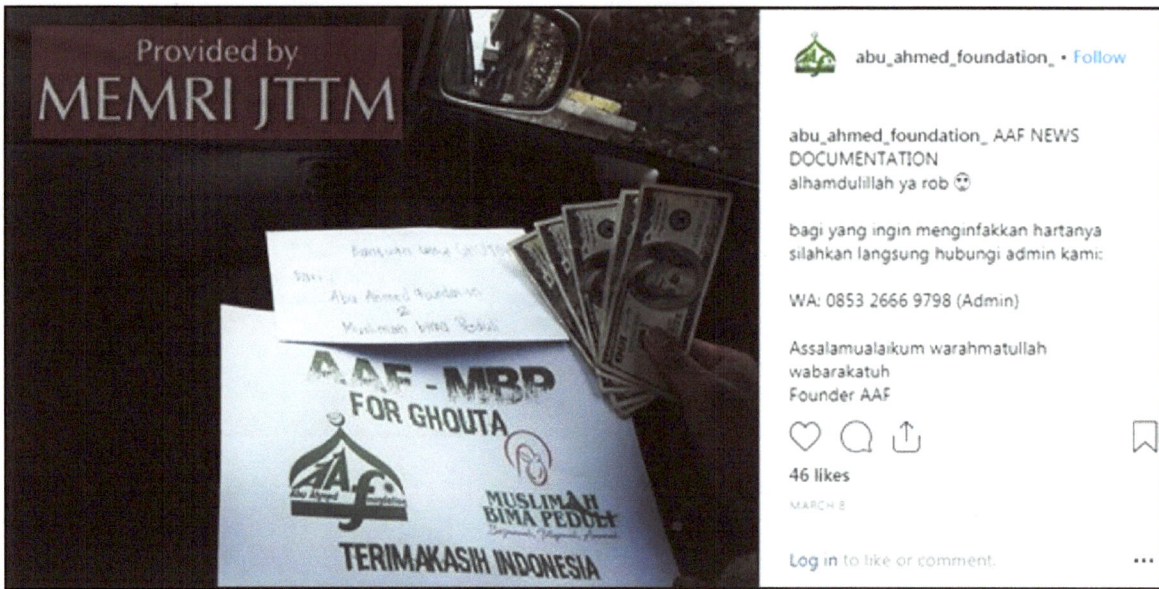

A July 17 post advocated for another project, "Teuku Umar," and included the same bank information and WhatsApp number.

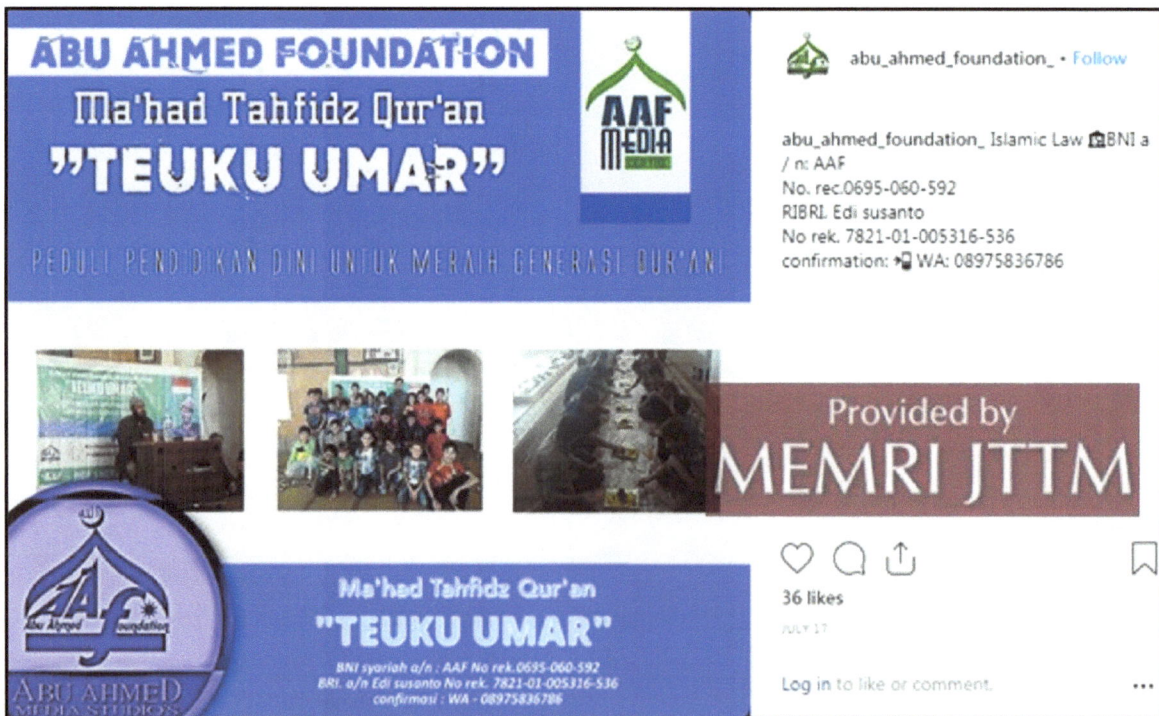

The Coming Storm: Terrorists Using Cryptocurrency

Pro-ISIS Gaza Jihadi Media Groups Ibn Taymiyya Media Center, Jaysh Al-Ummah Fundraise Using Cryptocurrency For Arming And Equipping Fighters In Gaza

Introduction

Beginning in 2015, the Gaza-based jihadi-Salafi media group Ibn Taymiyya Media Center (ITMC) and its Jahezona ("Equip Us") online campaign have been active online raising funds and using cryptocurrencies for jihadi activity such as purchasing weapons. The Jaysh Al-Ummah Al-Salafi, another group associated with Salafi-jihadi elements in Gaza, launched a similar "Equip A Fighter" campaign on social media. To encourage supporters to donate, the groups post, among other things, graphics giving the prices of various weapons on many social media platforms.

Ibn Taymiyya Media Center (ITMC) Launches Jahezona ("Equip Us") Fundraising Campaign To Arm Mujahideen in Gaza

On July 9, 2015, ITMC launched an online fundraising campaign on Twitter calling on Muslims to donate funds so the group could buy weapons and ammunition for the mujahideen in Gaza.[114]

يقوم على هذه الحملة ويُشرف عليها ثلة من المشايخ والمجاهدين نحسبهم والله حسيبهم من أهل السبق والصدق ممن أسسوا وقادوا ورعوا مسيرة التوحيد ودرب الجهاد في هذه البلاد منذ سنوات، فصبروا على الأذى وظلوا ثابتين رغم أمواج المحن والابتلاءات.

حملة جهزونا

للتواصل: تليغرام: @jahezona
تويتر: @jahezona02
ايميل: jahezona@tutanota.com

طامح الغزاوي @tameh_gaza2 · Jul 27

#حملة_جهزونا منكم المال ومنّا دمُنا مزكاة من مركز #ابن_تيمية_للإعلام تدعوكم لتنصروا دين الله تأملكم فلا تبحلوا

9   5

(حملة (جهزونا
@jahezona02

#حملة_جهزونا
لمن أراد التبرع وتجهيز المجاهدين في بيت المقدس إليكم أسعار السلاح الذي يحتاجه إخوانكم في جهادهم فساهموا.

View translation

طامح الغزاوي
@tameh_gaza2

مركز #ابن_تيمية_للإعلام يعلن انطلاق #حملة_جهزونا منكم المال ومنّا دمُنا لجمع التبرعات

justpaste.it/jahezona

#باقية

View translation

للتواصل: تويتر وتليغرام: @jahezona ايميل: jahezona@tutanota.com

RETWEETS FAVORITES
7   2

3:09 PM - 9 Jul 2015

The campaign, called "Equip Us," tweeted a graphic listing different weapons with their prices. The list gave the price of an RPG as $3,000, the price of a Kalashnikov rifle as $2,500, and the price of medium-size machine guns and sniper rifles as $5,500.

The list was accompanied by appeals for donations from those who do not wish to physically take part in jihad, encouraging them to carry out "financial jihad" by donating. Citing a hadith praising philanthropists, the campaign organizers made a special appeal to Muslim businessmen to donate generously.

The post listed contact information, including accounts on Telegram (@jahezona), and Twitter (@jahezona02), as well as an email address, jahezona@tutanota.com. The group had a backup Twitter account, @jahezona03, in case the primary one is suspended. The campaign's Twitter account had more than 2,000 followers at that time.

The Coming Storm: Terrorists Using Cryptocurrency

جهزونا@

تليجرام: @j a h e z o n a
تويتـر: @j a h e z o n a 0 2
احتياطي: @j a h e z o n a 0 3
إيميـل: jahezona@tutanota.com

منكـم المـال ومنـا دمُنـا...تابعونـا عبر هاشتاق: #حملة_

حملة لجمع التبرعات للمجاهدين على ثرى بيت المقدس بتزكية من مركز ابن تيمية

TWEETS	FOLLOWING	FOLLOWERS	FAVORITES
526	106	2,012	42

Tweets   Tweets & replies   Photos & videos

حملة (جهزُونا
@jahezona02

(حملة (جهزُونا
@jahezona02

jahezona@tutanota.com

bentymeia@

twitter.com/jahezona03

Tweet to   Message

حملة جهزُونا

لجمع التبرعات

منكم المال .. ومنا دمنا

ـحملة -جهزونا
@jahezona_014

#حملة_جهزونا..منكم المال ومنّا دمُنا::لدعم لمجاهدي غزة
مزكاة من مركز #ابن_تيمية_للإعلام.. فياباغي الخير أقبل
للتواصل تليجرام @jahezona
إيميل jahezona@tutanota.com

In December 2015,[115] the campaign's motto was "The money will come from you and the blood will come from us." The campaign stressed that the mujahideen were in dire need of such aid and that "waging jihad by means of money" (i.e., financially assisting the jihad fighters and their families) was an important religious duty incumbent upon every Muslim. The campaign emphasized that one who donates money for jihad was like one who fights with his own hands, and that the religious texts and scholars "even put this duty before" the duty of waging jihad on the battlefield. The campaign appealed to the Muslims' sentiments, calling upon them not to abandon the Gazan mujahideen in the struggle against "the enemies of the faith and those who violate our honor and our women's honor, [namely] the Jews," and harshly rebuked Muslims who failed to perform this duty. It reinforced its message by quoting leaders of global jihad who underscored the importance of donating money for the mujahideen, especially those in Palestine.

While initially the campaign was conducted mainly through a designated Twitter account (@ahezona_014), in December 2015 the campaign had opened a second account on Telegram (Telegram.me/jahezona02). The campaign's Twitter and Telegram were updated daily with relevant content and banners. Followers who wished to donate were instructed to contact the campaign operators via email (jahezona.@tutanota.com) or through the Telegram account.

The group posted a banner on Telegram on December 2, 2015. The banner gave the prices for a fighter's "personal weapon with ammunition" ($2,500), a "PK machine gun" ($5,500), "an [RPG grenade] launcher" ($3,000), and "a sniper's rifle" ($5,500). It stated: "Aside from these weapons, additional items [are needed] such as explosive charges and missiles. So assist your brothers, even with $10."

The campaign's main message was that jihad is a religious obligation that must be fulfilled either through actual fighting or by supporting the fighters, financially or otherwise. The campaign stressed that contributing to the war effort was equivalent to physically taking part in it. In announcing its campaign, the ITMC quoted a hadith conveying this message: "Whoever equips one who is fighting for the sake of Allah is like one who fights himself. Whoever provides for the family of a fighter... [during his absence] is also like one who fights himself..." The same hadith accompanied the photo below, posted on the campaign's Telegram account on November 25.

The Coming Storm: Terrorists Using Cryptocurrency

The ITMC announcement continued: "Your monotheist brothers on the periphery of Bayt Al-Maqdis [Jerusalem] need help. Their jihad has waned and their actions have dwindled due to lack of funds and [due to] economic difficulties. They are prevented from carrying out many jihad operations due to inability to finance them. Allah be praised, there are many men who strive to outdo one another in fervor and go to great lengths to fulfill Allah's will and draw closer to him. Lo, your brothers [in Gaza] go forth today to defend you and your religion; they have opened before you the gates to a reward so great that [the believers] should be vying with each other to earn it. So show Allah the good in you and help the jihad fighters with your bounty. Faced with difficult trials and intense isolation[116] on a daily basis, the mujahideen have nothing [to aid them] aside from Allah and your help. Know that whoever is stingy is being stingy only with himself, for Allah alone is wealthy and powerful."[117]

Another message, authored by a member of the Salafist-jihadi stream in Gaza and posted on Telegram by the head of the Equip Us campaign, further sharpens the message: "Whoever reads and studies the verses of the precious book [the Koran] and examines the Sunna of the faithful Prophet in depth... and whoever reads [the writings of] the ancient and modern religious scholars, discovers that the texts calling to [wage] jihad and invest in it are innumerable. The holy books are equally replete with evidence regarding the virtue and importance of jihad by means of money. [In fact, they put this kind of jihad] even before jihad with one's life... The importance and great virtue of this commandment are well-known; it is known to be the very nerve, heart and body of jihad. Through it, Allah's religion will be established and without it the embers of jihad will be extinguished. Woe to a nation that spends much more on pleasures than on repelling the aggressor and keeping the fuse of jihad burning bright to thwart the [enemy's] plot and rebuff the oppressor and liberate what was stolen by the enemies of Allah and the nation. Woe to it!"[118]

حملة جهزونا
لجمع التبرعات
لإخوانكم المجاهدين على ثرى بيت المقدس

حملة (جهِّزُونا).. منكم المال ومنّا دمُنا

حملة لجمع التبرعات للمجاهدين على ثرى بيت المقدس بتزكية من مركز ابن تيمية للإعلام

بسم اللَّه الرحمن الرحيم

الحمد للَّه رب العالمين والصلاة والسلام على نبينا محمد وآله وصحبه أجمعين .

وبعد ....

إن القارئ المتأمل لأي الكتاب العزيز والمقلب في سنة النبي الأمين –صلى اللَّه عليه وسلم– وكذا القارئ لأهل العلم في القديم والحديث؛ ليجد أنّ النصوص المحرضة على الجهاد وبذل الوسع فيه لا حصر لها ولا عدّ، وكذا تخصيص الجهاد بالمال قبل النفس طافحة النصوص تدليلاً وتحريضاً وتنبيهاً على فضائله وخطورة أمره، وعظيم التقصير فيه وبه !

قال – تعالى –: (انفِرُوا خِفَافًا وَثِقَالًا وَجَاهِدُوا بِأَمْوَالِكُمْ وَأَنْفُسِكُمْ فِي سَبِيلِ اللَّهِ ذَلِكُمْ خَيْرٌ لَكُمْ إِنْ كُنْتُمْ تَعْلَمُونَ) [التوبة: ٤١] و قال : (يَا أَيُّهَا الَّذِينَ آمَنُوا هَلْ أَدُلُّكُمْ عَلَى تِجَارَةٍ تُنْجِيكُمْ مِنْ عَذَابٍ أَلِيمٍ * تُؤْمِنُونَ بِاللَّهِ وَرَسُولِهِ وَتُجَاهِدُونَ فِي سَبِيلِ اللَّهِ بِأَمْوَالِكُمْ وَأَنْفُسِكُمْ ذَلِكُمْ خَيْرٌ لَكُمْ إِنْ كُنْتُمْ تَعْلَمُونَ) [الصف: ١٠-١١] وقال –جل وعلا– : (لَا يَسْتَوِي الْقَاعِدُونَ مِنَ الْمُؤْمِنِينَ غَيْرُ أُولِي الضَّرَرِ وَالْمُجَاهِدُونَ فِي سَبِيلِ اللَّهِ بِأَمْوَالِهِمْ وَأَنْفُسِهِمْ فَضَّلَ اللَّهُ الْمُجَاهِدِينَ بِأَمْوَالِهِمْ وَأَنْفُسِهِمْ عَلَى الْقَاعِدِينَ دَرَجَةً وَكُلًّا وَعَدَ اللَّهُ الْحُسْنَى وَفَضَّلَ اللَّهُ الْمُجَاهِدِينَ عَلَى الْقَاعِدِينَ أَجْرًا عَظِيمًا) [النساء : ٥٩]

قال الآلوسي – رحمه اللَّه –: «لعل تقديم الأموال على الأنفس لـما أن المجاهدة بالأموال أكثر وقوعاً، وأتم دفعاً للحاجة؛ حيث لا يُتصوَّر المجاهدة بالنفس بلا مجاهدة بالمال، وقيل: ترتيب هذه المتعاطفات في الآية على حسب الوقوع؛ فالجهاد بـ(المال) لنحو التأهب للحرب، ثم الجهاد بالنفس»

وقال صاحب البرهان : «وجه التقديم أن الجهاد يستدعي تقديم إنفاق الأموال أولاً فهو من باب السبق بالسببية»

قلت : صدق – رحمه اللَّه – فإن دفع العدو الصائل واجبٌ ولا يتم هذا الواجب إلا بالتجهز له والإعداد، وهذا الأخير مفتقرٌ إلى بدل ومال، وقد حكى الفقهاء أن : ما لا يتم الواجب إلا به فهو واجب .

قال ابن القيم – رحمه اللَّه – في حكمة تقديم المال على النفس:

أولاً: هذا دليل على وجوب الجهاد بالمال كما يجب بالنفس، فإذا دهم العدو وجب على القادر الخروج بنفسه، فإن كان عاجزاً وجب عليه أن يكتري بماله.

ومن تأمل أحوال النبي – صلى اللَّه عليه وسلم – وسيرته في أصحابه – رضي اللَّه عنهم – وأمرهم بإخراج أموالهم في الجهاد، قطع بصحة هذا القول .

والمقصود: تقديم المال في الذكر، وأن ذلك مشعرٌ بإنكار وهم مَنْ يتوهم أن العاجز بنفسه إذا كان قادراً على أن يغزو بماله لا يجب عليه شيء؛ فحيث ذكر الجهاد قدَّم ذكر المال؛ فكيف يقال: لا يجيب به؟

ولو قيل: إن وجوبه بالمال أعظم وأقوى من وجوبه بالنفس، لكان هذا القول أصح من قول من قال: لا يجب بالمال، وهذا بيِّن، وعلى هذا فتظهر الفائدة في تقديمه في الذكر.

The Coming Storm: Terrorists Using Cryptocurrency

A Telegram post from December 2, 2015, emphasized that even small donations or actions of publicizing the campaign, are of great benefit: "Be the one who enabled the mujahid to equip himself [with weapons] while you sit at home. Donate even a [small] sum, enough for a few bullets in the fighter's magazine clip, or else spread the word about this [Telegram] account, and thus make sure that our words reach the donors."[119]

The campaign stressed that the act of donating will flood the donor with a sense of pure joy. A banner posted on Twitter on November 24 said: "[Here is] a practical experiment. Donate 10 dinars to your brother and see what happiness and pure joy you feel."[120]

The campaign declared that the funds are intended for jihad in Palestine and played extensively on anti-Jewish sentiment. For example, a banner posted on Telegram on December 4, 2015, showed the Dome of the Rock in Jerusalem and promised: "Soon we will burn the Jews, and upon our severed limbs victory shall be established."

A December 4, 2015 post on Telegram read: "Our brother! Does it not gladden you when you see us killing the enemies of the faith and those who violate our honor and our women's honor – the Jews!? Why not start [the fight against them] yourself by donating, because giving money precedes giving blood?"

Another frequent theme of the campaign was rebuking Muslims who do nothing to assist the Gaza mujahideen. A December 1, 2015 post on Twitter read: "It is as though some of our brothers are not interested in our cause. It is as though we are not their brothers. It is as though Allah does not obligate you to assist those who support you in religion."[121]

A banner posted November 30 on Twitter read: "Muslims seem to identify deeply with the suffering and problems of their brothers in Bayt Al-Maqdis. So why do we not see you translating this solidarity into aid for us??? Nothing is more beautiful than a mujahid in Bayt Al-Maqdis targeting the fortifications of the Jews with a missile you paid for, while you sit at home."

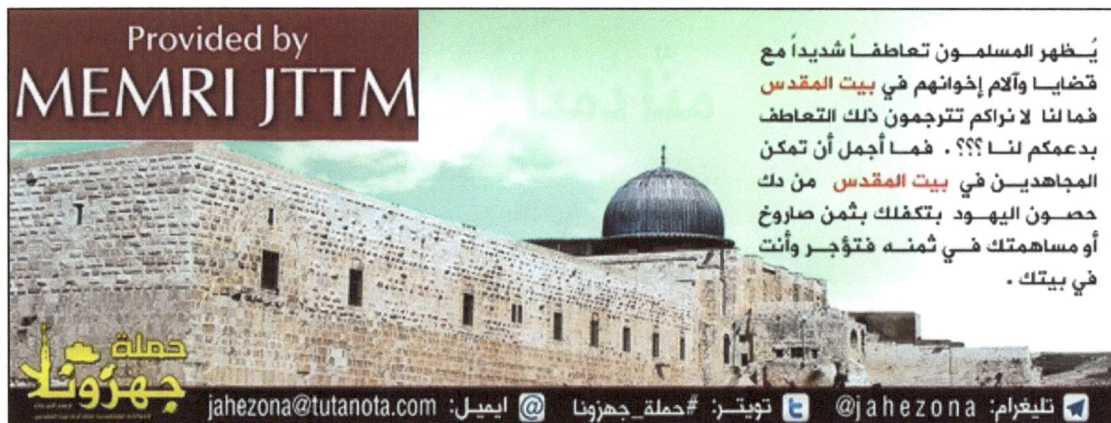

The Coming Storm: Terrorists Using Cryptocurrency

Though the campaign raises funds for jihad in Gaza, some of its aspects reflect its ideological affiliation with global jihad, including photos of Gazan fighters who were killed on jihad fronts outside of Palestine. A banner posted November 29, 2015 on Telegram showed Mahmoud Al-Salfiti, who was jailed by Hamas for killing an Italian activist, Vittorio Arrigoni, in April 2011 but later escaped and was killed fighting with ISIS in Iraq. The banner called him "one of the knights of the Islamic Caliphate State, from Gaza, [who was] killed in battles between those loyal to [Allah] the Compassionate and the satanic soldiers of the Safavid [i.e. Shi'ite] army."[122]

The campaign also frequently quoted prominent commanders of global jihad regarding the duty of every Muslim to assist the mujahideen, especially those in Palestine. For example, a message posted on Telegram on November 26 quoted Osama bin Laden as saying: "I urge the young men and the faithful merchants to seize this golden opportunity to fulfil the great duty of defending the faith and saving the ummah, and this by supporting the jihad with money, by inspiring [others to fight], and by fighting our enemies – especially in Palestine and Iraq."[123]

Another post quoted Abu Omar Al-Baghdadi, the former leader of the Islamic State of Iraq, a predecessor to ISIS. He suggested that "every Muslim donate two dollars a month out of his earnings: one dollar for our people in Palestine and another for the rest of the [jihad] fronts." He also suggested the forming of secret organizations to forward these funds.

In June 2016, the ITMC "Equip Us" campaign began soliciting donations in bitcoin.[124]

A poster soliciting donations for the mujahideen in Gaza.

A poster from the campaign pricing the "cost for equipping a mujahid" at $2,500

The group's first solicitation for donations using Bitcoin was posted on June 29, 2016, on the campaign's Telegram account: "In addition to the secure means through which we receive your donations, you can send donations via #bitcoin using the attached [bitcoin wallet address] code." The bitcoin address given was 1MMaU5nTrFdPZotfwdbv1wWnFjLCTFbpPY.

The Coming Storm: Terrorists Using Cryptocurrency

Provided by
MEMRI JTTM

إضافة إلى الوسائل الآمنة التي نستقبل بها تبرعاتكم

بإمكانكم إرسال التبرعات عبر #البتكوين باستخدام الكود المرفق

رقم المحفظة :

1MMaU5nTrFdPZotfwdbv1wWnFjLCTFbpPY

Provided by
MEMRI JTTM

A pricelist for various rockets showing their "effectiveness" in targeting Jews and Jewish settlements, along with the bitcoin address QR Code.

A blockchain.info lookup of that address around that time showed a single transaction, on July 1, for $555.59.

In March 2017, the ITMC "Equip Us" campaign[125] continued to solicit donations on Telegram.[126] One poster urged supporters to donate before it's "too late." Weapon prices appeared in USD on the poster.[127]

The Coming Storm: Terrorists Using Cryptocurrency

"Patience, oh Aqsa, for the nation of Muhammad... continues to be fertile with mujahideen [who sacrifice] themselves and their money for the path of Allah. And as the lions of sacrifice gave their souls generously, the lions of giving and donation will give their money generously to equip the mujahideen."[128]

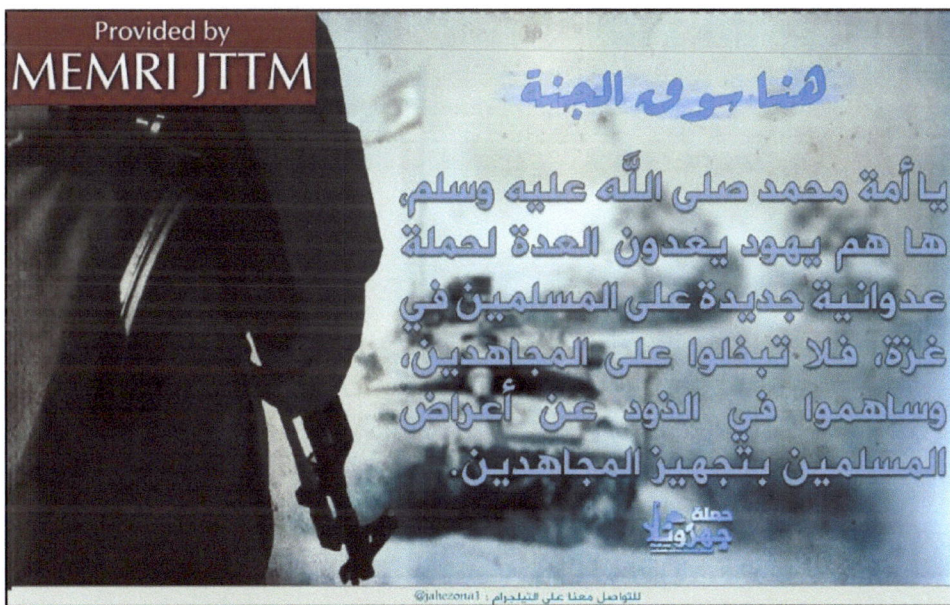

"Here is the market of Paradise. Oh nation of Muhammad... here they are, the Jews preparing [to launch] a new aggression campaign on the Muslims in Gaza, so do not be stingy with the mujahideen, and take part in protecting the honor of Muslims by equipping the mujahideen."[129]

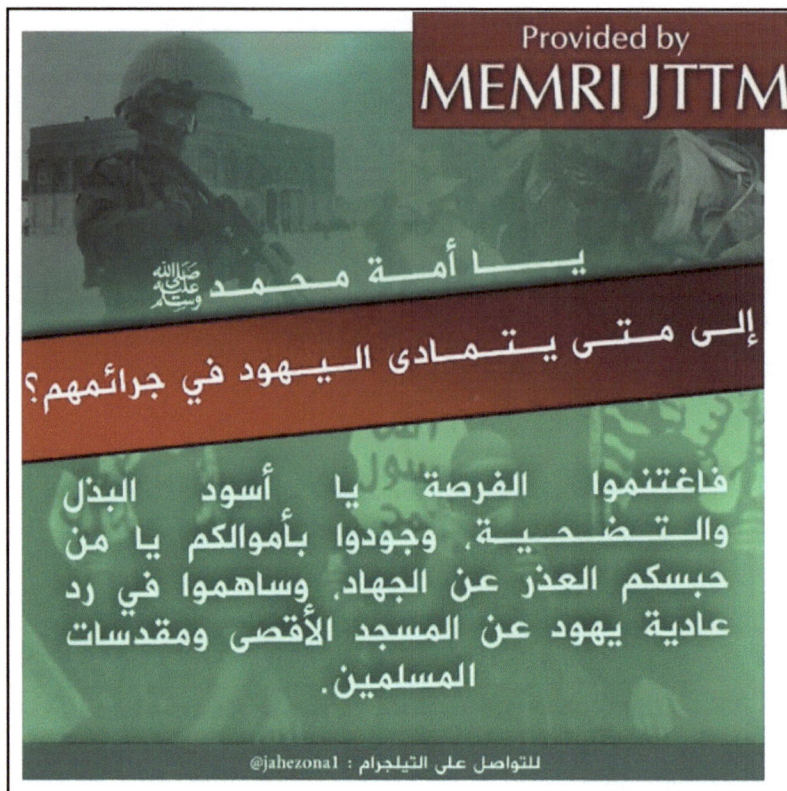

يا أمة محمد ﷺ

إلى متى يتمادى اليهود في جرائمهم؟

فاغتنموا الفرصة يا أسود البذل والتضحية. وجودوا بأموالكم يا من حبسكم العذر عن الجهاد. وساهموا في رد عادية يهود عن المسجد الأقصى ومقدسات المسلمين.

للتواصل على التيلجرام : @jahezona1

"Oh nation of Muhammad... how much longer will the Jews continue with their crimes? Seize the moment, oh lions of giving and sacrifice, and give your money generously, oh you who were excused from [joining the] jihad. Take part in responding to the Jews' offense against the Al-Aqsa Mosque and the holy places to Muslims."[130]

أيّها المُقتدرون

ماذا لو تبرعتم لإخوانكم المجاهدين بثمن صاروخ في كل شهر أيعجزكم ذلك أم أنّكم زهدتُم في هذا الأجر يا كرام!

للتواصل معنا #حملة_جهزونا

تليغرام @jahezona1

"Oh you [who are] capable, what if you donated to your mujahideen brothers the price of a missile every month, would that cripple you [financial-ly]?..."[131]

The Coming Storm: Terrorists Using Cryptocurrency

حملة جهزونا

(وأعدوا لهم ما استطعتم من قوة، ألا إن القوة الرمي)

صاروخ كتيوشا 107 - دولي	$950
المدى	8 Km
يهود تحت النيران	31,590
عدد المستوطنات	21
عدد المدن الكبرى	1
القوة التدميرية	متوسطة
دقة الإصابة	95%

صاروخ كتيوشا 107 - محلي	$200
المدى	4.5 Km
يهود تحت النيران	6,083
عدد المستوطنات	15
عدد المدن الكبرى	0
القوة التدميرية	متوسطة
دقة الإصابة	75%

صاروخ "جراد" 40 - دولي	$5000
المدى	42Km
يهود تحت النيران	574,072
عدد المستوطنات	20+
عدد المدن الكبرى	7
القوة التدميرية	كبيرة
دقة الإصابة	85%

صاروخ "جراد" 20 - دولي	$2700
المدى	21 Km
يهود تحت النيران	155,477
عدد المستوطنات	20+
عدد المدن الكبرى	3
القوة التدميرية	كبيرة
دقة الإصابة	90%

للتواصل: تليجرام: @jahezona1 | تويتر: #حملة_جهزونا @jahezona1 | إيميل: jahezona@tutanota.com

Missiles, their prices, and various related stats.

On June 5, 2017, the ITMC "Equip Us" campaign posted an infographic on its Telegram channel[132] highlighting the allocation of donations received thus far.[133] Seven of the eight categories listed in the infographic pertained to jihad-related activities, including: military training; manufacture of weapons, including missiles and IEDs, and the training of the next generation of weapons manufacturers; arming the mujahideen; participation in jihad operations; jihadi media, including the training of a jihadi media cadre; social services for the mujahideen, for the families of martyrs and prisoners, and for poor Muslims in Gaza; and various activities aimed at improving the operational security of the mujahideen.

The Gaza-Based Jaysh Al-Ummah Jihadi Group's Annual 'Equip a Fighter' Fundraising Campaign: Urging Muslims To Donate Using Bitcoin During Ramadan

Since March 2015, the Gaza-based jihadi-Salafi Jaysh Al-Ummah Al-Salafi group, which is associated with jihadi elements in Gaza, has run a near-annual "Equip A Fighter" fundraising campaign urging Muslims around the world to donate to equip its fighters and support the jihad in Palestine. The campaign has been promoted across social media platforms.

The 2017 version of the campaign was kicked off on March 8 on Twitter, Telegram, and WhatsApp.

Billed in the announcement by the group's media arm, Al-Raya, as raising funds to "counter the usurper Jews and implement Allah's sharia,"[134] the "motives and reasons" behind the campaign were to "raise the word of monotheism in Palestine, implement Allah's rules, support the oppressed, counter the aggression of the infidels, liberate the sacred places, sponsor the orphans and the families of martyrs, and support the brave prisoners and wounded fighters."

The Coming Storm: Terrorists Using Cryptocurrency

The campaign organizers provided multiple means of communication for donors including a phone number for Skype and WhatsApp, a Twitter account, a Telegram channel, and a Mail.com email address.

On Twitter, the campaign organizers dedicated accounts and hashtags to promoting the campaign and encouraging Muslims to donate, posting statements and graphics featuring quotes by prominent Islamist ideologues and jihadi figures highlighting the importance of providing financial supports to jihadi groups. In addition to publishing these statements, the organizers used posts that include names, images, and prices of various weapons to encourage people to donate.

Poster showing prices of weapons and equipment that donations will be used to purchase.

In May and June 2018, the group intensified its campaign on Telegram, Facebook, Twitter, Instagram, Google Plus, LinkedIn, Skype, WhatsApp, and WordPress.[135]

In one of its posters on Telegram, the group urged Muslims to "contribute with us even if by a little to equip a mujahid for the sake of Allah and supply him with weapons and equipment during this blockade that has been imposed on the mujahideen who need what enables them to wage jihad." According to the poster, equipping one fighter costs $2,000.[136]

Another poster on these platforms listed different types of weapons, including rockets, and their prices, for those interested in having their donation be used to buy a specific weapon.

The group also used hashtags such as "Equip A Fighter," "Equip Us Campaign," and "Jaysh Al-Ummah Al-Salafi," and asked readers to share these hashtags along with the new accounts and backup accounts in case the accounts were suspended.

Those interested in donating to the campaign were asked to contact the organizers by using the email address omma.ps@mail.com, the Skype username Mohjat Al-Quds ("Soul of Jerusalem"), the WhatsApp number 972592120765, or the Telegram channel @Omma_ps.

In 2018, although the campaign organizers did not specify how funds could be transferred to them, at least one communication informed donors that donations could be sent using bitcoin, and provided an address on the Blockchain website: a41d56d7-ec3a-4747-a0c2-07cfd7046a32.

The Coming Storm: Terrorists Using Cryptocurrency

Google+
Sign in
Discover
Join Google+
Report an issue
Help
دعـــوة
لبذل المال في سبيل الله
في زمن عز فيه النصير
وكثر فيه الخذلان
جهز غازيا
التابعة لجيش الأمة السلفي في بيت المقدس
Provided by
MEMRI JTTM
جهز غازياً
ABOUT

دعـــوة
حملة
Provided by
MEMRI JTTM
جهز غازيا
التابعة لجيش الأمة السلفي في بيت المقدس
Tweets
8
Followers
4
Follow
جهز غازياً #جهزونا
@OmmaPs2
Tweets
Tweets & replies
Media
New to Twitter?
Sign up

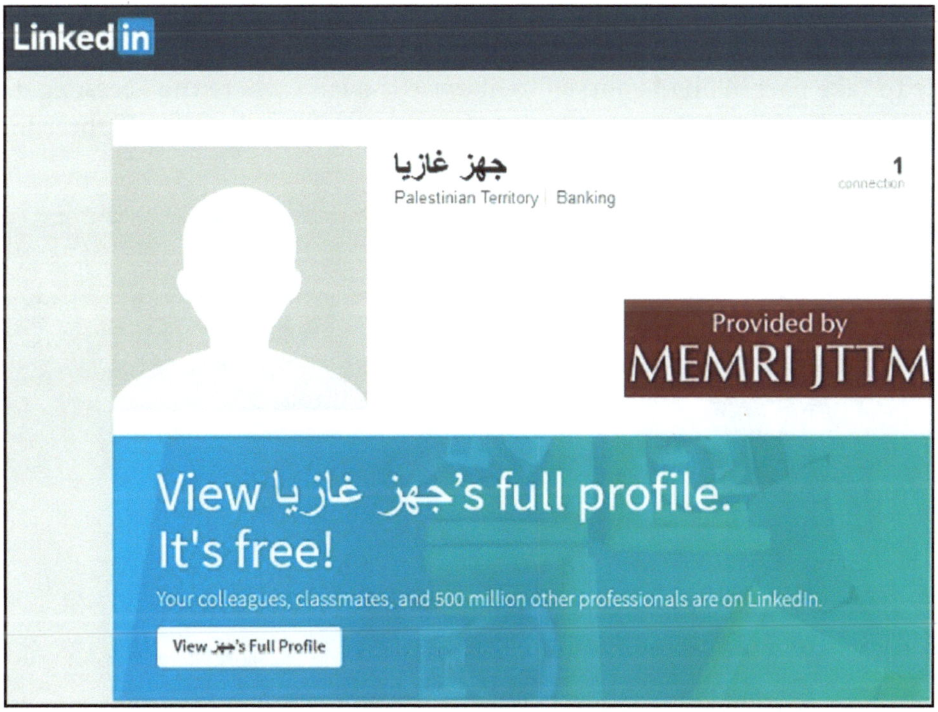
LinkedIn
جهز غازيا
Palestinian Territory | Banking
1
connection
Provided by
MEMRI JTTM
View جهز غازيا's full profile.
It's free!
Your colleagues, classmates, and 500 million other professionals are on LinkedIn.
View جهز's Full Profile

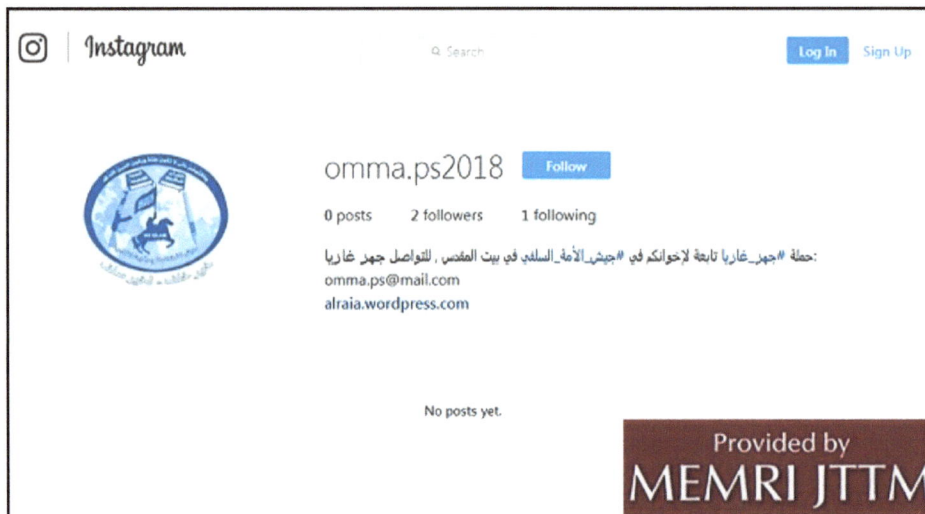

On May 7, 2019, Jaysh Al-Ummah's media arm Al-Raya relaunched the "Equip A Fighter" campaign – this time specifically calling on Muslims to donate using bitcoin during the month of Ramadan. The fundraising poster, which was published on Al-Raya's Telegram channel, noted that the campaign would last 30 days and provided contact information, by email at Omma.ps@mail.com and on Telegram at @Omma_ps. It also included the campaign's address on the Blockchain website, with the same Bitcoin address as in the previous year: a41d56d7-ec3a-4747-a0c2-07cfd7046a32.

The poster called on Muslims to "donate money for the sake of Allah to support the mujahideen in Jerusalem in their jihad to defeat the Jews and stop their aggression." It added: " If you are unable to pray at Al-Aqsa or provide oil [for the lamps], here is our blood that we would gladly sacrifice for the sake of Allah and to liberate the Al-Aqsa mosque. If you are unable to take part in jihad yourself to liberate the land where the Messenger of Allah's night journey took place then wage jihad with your wealth and equip a fighter for the sake of Allah."[137]

The Coming Storm: Terrorists Using Cryptocurrency

On May 14, 2019, the group extended the campaign to other social media platforms, including Twitter, Instagram, and LinkedIn, and urged its followers to promote it using the hashtags "Equip Us" and "Equip A Fighter." On Al-Raya's Telegram channel and WordPress page, the group asked followers unable to send money to share the campaign's announcement across social media platforms. The campaign announcements included posters showing different types of weapons and their prices, as well as quotes from senior jihadi figures encouraging Muslims to donate to support the jihad and the mujahideen, and gave an email address, Omma.ps @mail.com.[138]

A few days later, on May 20, 2019, Jaysh Al-Ummah released a short, 80-second promotional video about the "Equip A Fighter" fundraising campaign, on the Al-Raya Telegram channel. The video presented the campaign as raising funds to "equip your brothers the mujahideen in the Salafi Jaysh Al-Ummah in Jerusalem to fight the usurper Jews and implement Allah's shari'a." It stated that the "motives and reasons" behind the "Equip A Fighter" campaign are to "raise the word of monotheism in Palestine and implement Allah's shari'a on Allah's land; to support the oppressed and counter the belligerence of the unbelievers; to liberate the sacred places and support the people of monotheism in Palestine; to support the call of Ahl Al-Sunnah Wal-Jama'ah [i.e. Sunnis] in confronting the groups of falsehood and deviation... to sponsor the orphans and the families of martyrs; and to support the heroic detainees and wounded [fighters]." The video reiterated the multiple ways donors could contact the campaign, including a Mail.com email address at Omma.ps@mail.com and the Telegram handle @Omma_ps, and a Bitcoin address, 1EM4e8eu2S2RQrbS8C6aYnunWpkAwQ8GtG.[139]

The following are images from the video.

The Coming Storm: Terrorists Using Cryptocurrency

TRANSLATED BY MEMRI TV

حملة جهّز غازياً : هي حملة لبذل الأموال في سبيل الله ولتجهيز إخوانكم المجاهدين في جيش الأمة السلفي في بيت المقدس ؛ لمقارعة اليهود الغاصبين وتحكيم شريعة ربّ العالمين

in order to equip your *mujahideen* brothers

TRANSLATED BY MEMRI TV

حملة جهّز غازياً : هي حملة لبذل الأموال في سبيل الله ولتجهيز إخوانكم المجاهدين في جيش الأمة السلفي في بيت المقدس ؛ لمقارعة اليهود الغاصبين وتحكيم شريعة ربّ العالمين

in the Salafi Jaysh Al-Ummah in Jerusalem with weapons and equipment,

TRANSLATED BY MEMRI TV

حملة **جهّز غازياً** : هي حملة لبذل الأموال في سبيل الله ولتجهيز إخوانكم المجاهدين في جيش الأمة السلفي في **بيت المقدس** ؛ لمقارعة اليهود الغاصبين وتحكيم شريعة ربّ العالمين

so they can fight the usurper Jews and implement Allah's *shari'a*.

TRANSLATED BY MEMRI TV

للتواصل معنا:

omma.ps@mail.com

@Omma_ps

1EM4e8eu2S2RQrbS8C6aYnunWpkAwQ8GtG

Also on May 20, Jaysh Al-Ummah released another video as part of its "Equip a Fighter" campaign. The video featured a group of fighters; one of them addressed Muslims around the world in English, urging them to support the mujahedeen monetarily as part of the ongoing fundraising campaign. It also included scenes of rocket shelling from multiple locations, and again provided the campaign's previously published contact information and bitcoin address.

Additionally, the next day, May 21, 2019, the Al-Raya Telegram channel shared a poster showing various types of "light weapons" and their prices in U.S. dollars, along with a call for Muslims to fund their purchase. According to the poster, a Kalashnikov rifle costs $1,700; a Dragunov sniper rifle costs $7,000; a "BKS machine gun," probably a PK machine gun, costs $5,000; and an "RBG," an RPG launcher, costs $2,500. The poster also gave the campaign's contact information and bitcoin address.[140]

The Coming Storm: Terrorists Using Cryptocurrency

May 2019: Al-Qaeda-Affiliate "Incite The Believers" In Syria Launches An "Equip Us" Fundraising Campaign On Telegram And WhatsApp

On May 18, 2019, the Incite the Believers operations room, a confederation of jihadi factions that includes Hurras Al-Din, an Al-Qaeda affiliate operating in Syria, launched a fundraising campaign on Telegram and WhatsApp under the name Jahezona ("Equip Us"), with an announcement on its Telegram channel stating that its aim was to "support the battlefronts financially" and that it is under the direct supervision of the operations room. It added that those interested in donating to the campaign should contact the group via Telegram @jahizona or WhatsApp at 963-931652475. A link to the campaign's WhatsApp group was also posted: Chat.whatsapp.com/IDcvlgR3Mqp3ZAwODD8DCP.

The campaign published an infographic showing statistics about the operations room's activities since its establishment in November 2018: over 110 operations, including 11 raids, three "special operations," more than 34 sniper attacks and 62 attacks with heavy artillery; more than 348 enemy soldiers killed, and more than 93 wounded. On May 19, the channel published a poster with an "urgent appeal" to all Muslims, writing that as the Assad regime "fights, bombs, and displaces" the Syrian people and the mujahideen hold off the regime and prepare for assault, the Incite the Believers operations room is "in the most dire need of funds" to continue its defense of "the liberated Sunni areas" and begin an offensive against the regime and its allies. The funds, stated the poster, were required for the operations room's current defense against the attacks of "the infidel regime;" for ensuring that the group's fighters on the front lines have bunkers, trenches, supply lines, and ammunition; for funding operations behind enemy lines; and for providing fighters with "logistical services," such as fuel, food, medical treatment, and maintenance of their equipment and vehicles.[141]

حملة جهزونا – غرفة عمليات وحرض المؤمنين

قال تعالى [إِنَّمَا الْمُؤْمِنُونَ الَّذِينَ آمَنُوا بِاللَّهِ وَرَسُولِهِ ثُمَّ لَمْ يَرْتَابُوا وَجَاهَدُوا بِأَمْوَالِهِمْ وَأَنْفُسِهِمْ فِي سَبِيلِ اللَّهِ أُولَئِكَ هُمُ الصَّادِقُونَ]

نداء عاجل

إلى جميع المسلمين النظام يقاتل ويقصف ويهجر، والمجاهدون يصدون ويعدون العدة للهجوم.

وتقوم غرفة عمليات وحرض المؤمنين العسكرية بحملة تجهيز للدفاع عن مناطق اهل السنة المحررة وذلك لصد الهجوم الهمجي لقوات النظام الكافر المجرم وأحلافه، ثم الأخذ بزمام المبادرة وتحويل الدفاع إلى هجوم بالتعاون مع الصادقين.

ولا يزال المجاهدون وخاصة غرفة عمليات وحرض المؤمنين يقومون بالغارة تلو الغارة بروح قتالية عالية ولله الحمد.

أهداف الحملة:

1- تمويل حملة ... النظام الكافر.
2- تجهيز خطو ... الذخيره والعتاد).
3- العمل خلف خطوط العدو وتجهيز المقاتلين وتفريغهم للقتال.
4- تجهيز الخدمات اللوجستية من وقود وطعام وعلاج وصيانة معدات وآليات عسكرية.

أيها المسلم ما دورك:

1- الخروج للقتال والرباط والتحصين لمن يتعين عليه الجهاد
2- الجهاد بالمال وتجهيز المقاتلين وخطوط القتال والرباط والمواجهة.

واعلم أن المجاهدين اليوم في أمس الحاجة للمال فإن المال عصب الجهاد وأنه بدون أموال ستضعف قدرة المجاهدين.

للدعم والمشاركة يرجى التواصل على الأرقام التالية :
00963931652475
@jahizona

أو صناديق التبرعات الخاصة بالحملة في المساجد.

حملة جهزونا

غرفة عمليات وحرض المؤمنين

The Coming Storm: Terrorists Using Cryptocurrency

Hamas's Al-Qassam Brigades Promote Bitcoin Use, Solicit Bitcoin Donations Worldwide – And Announces Plans To Send "Hundreds Of Millions" Of Text Messages In Multiple Languages Explaining How To Donate

Hamas Military Wing Al-Qassam Brigades Calls On Supporters Worldwide To Send Funds Using Bitcoin

On January 29, 2019, Hamas's military wing, the Izz Al-Din Al-Qassam Brigades, called on its supporters to send it funds in Bitcoin. It should be noted that the Al-Qassam Brigades are listed as a terrorist organization by the E.U.,[142] the U.S.,[143] Australia,[144] New Zealand,[145] and the U.K.[146] The fundraising campaign was disseminated widely, across many social media platforms, including Telegram, Facebook, Twitter, Instagram, Google Plus, LinkedIn, Skype, WhatsApp, and WordPress.[147]

Al-Qassam's announcement, which was posted on the Telegram channel of Al-Qassam spokesman Abu Ubaida, hinted that Bitcoin would be a way to combat Israel's crackdown on aid for the group. The message read: "The Zionist enemy combats the resistance [i.e. Hamas] by attempting to stop the support [it receives] by any means, but the supporters of the resistance in the entire world are fighting these Zionist's attempts and are seeking to find all possible means to support it... We call upon all supporters of the resistance and aides of our just cause to support the resistance financially through Bitcoin using mechanisms that we will soon publish."[148]

Provided by
MEMRI JTTM

"أبو عبيدة" الناطق العسكري باسم كتائب القسام"

● أبو عبيدة: يحارب العدو الصهيوني المقاومة من خلال محاولة قطع الدعم عنها بكل السبل لكن محبي المقاومة في كل العالم يحاربون هذه المحاولات الصهيونية ويسعون لإيجاد كافة سبل الدعم الممكنة للمقاومة.
👁 5413 1:42 PM

"أبو عبيدة" الناطق العسكري باسم كتائب القسام"

● أبو عبيدة: ندعو كل محبي المقاومة وداعمي قضينا العادلة لدعم المقاومة ماليا من خلال عملة ال "بيتكوين" عبر الآليات التي سنعلن عنها قريباً
👁 5243 1:44 PM

Also on January 29, 2019, Abu Ubaida wrote on Instagram: "The Zionist enemy fights the Palestinian resistance by trying to cut aid to the resistance by all means, but lovers of resistance around the world fight these Zionist attempts and seek all possible means to aid the resistance." He promised to supply more details later of how supporters could contribute by Bitcoin.[149]

Gaza Academic And Journalist Hussam Al-Dajany: People Can Now Donate To Hamas Undetected Using Bitcoin

On January 30, 2019, Palestinian journalist Hussam Al-Dajany, who also lectures at Gaza Open University, said in an interview on Hamas's Al-Aqsa TV that Hamas is accepting Bitcoin donations. Bitcoin, he explained, provides an opportunity to unite people who are afraid to or cannot openly support the Palestinian resistance because it is safe and reliable, and it cannot be tracked by security agencies. He added that money is what the resistance needs the most and that it costs millions of dollars to pay salaries, manufacture rockets, smuggle weapons, dig tunnels, and build drones. Despite Iran's open support of the Palestinian resistance, he said, the sanctions imposed on it might prevent enough money from getting to Hamas, so Bitcoin is being resorted to as an indirect way for people to "participate in the liberation of the Al-Aqsa Mosque."[150]

Hamas Military Wing Al-Qassam Brigades Publishes Its Bitcoin Address To Accept Donations

On January 31, 2019, the Al-Qassam Brigades published its Bitcoin address and announced it was accepting donations. The announcement came a day after an Al-Qassam Brigades' spokesman revealed that the group would be using the cryptocurrency to combat Israel's constraints on the group's funding. This announcement too was posted on the Al-Qassam Brigades' Telegram channel. The address shows eight transactions with about $110 in funds received in Bitcoin (BTC) and less than a dollar in Bitcoin Cash (BCH).

194

الإعلام العسكري
www.alqassam.net

تعلن

كتائب الشهيد عز الدين القسام

عن بدء استقبال

دعمكم المالي للمقاومة

بعملة الـ "بيتكوين"

عبر عنوان المحفظة الرسمي والوحيد:

3PajPWymUexhewHPczmLQ8CMYatKAGNj3y

يرجى نسخ عنوان المحفظة

من الموقع الرسمي لكتائب القسام

www.alqassam.net

Al-Qassam Brigades Updates Bitcoin Address For Greater Privacy And Security

On February 2, 2019, Al-Qassam Brigades updated its bitcoin address, stating that this was in order to increase privacy and security. The new bitcoin address is 17QAWGVpFV4g-Z25NQug46e5mBho4uDP6MD.[151]

The Coming Storm: Terrorists Using Cryptocurrency

Al-Qassam Brigades Publishes New Posters Promoting Bitcoin Fundraising Campaign

On February 4, 2019, the Al-Qassam Brigades published new posters promoting its Bitcoin fundraising campaign. The account has so far received $900.[152]

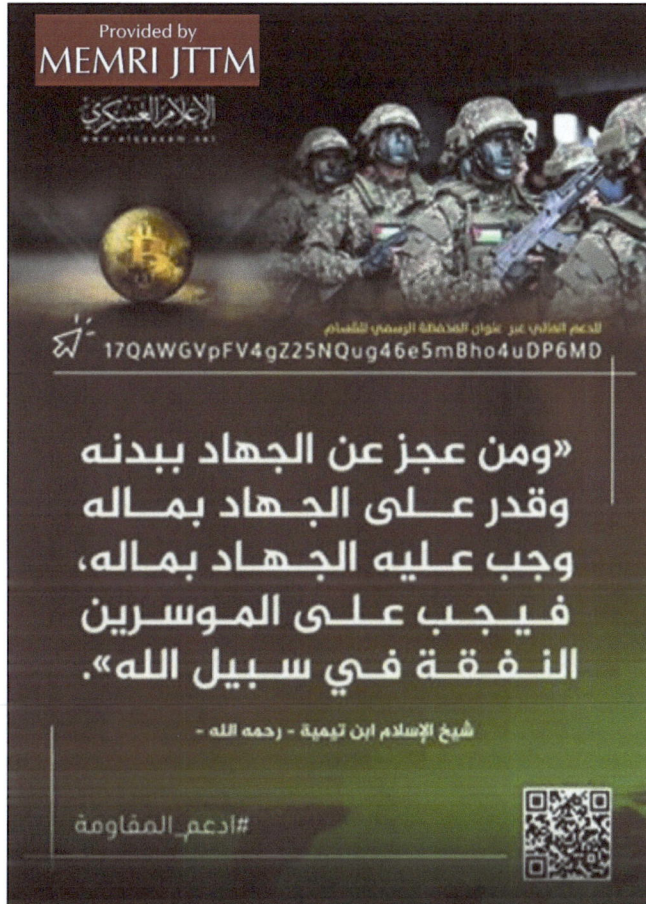

Alqassam.net/arabic, accessed February 4, 2019

Alqassam.net/arabic, accessed February 4, 2019

Stalinsky ■ MEMRI ميمري

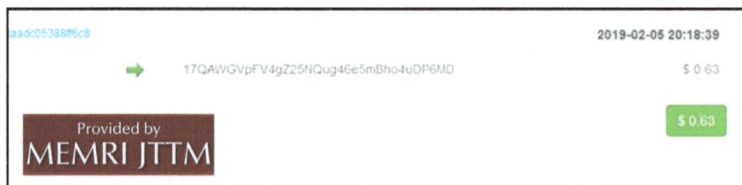

On February 5, 2019, Al-Qassam's latest Bitcoin address was still active, showing nearly $2,000 in total received funds. The blockchain also showed very small sums (for example, the 63 cents below) – a practice known as "dusting" aimed generally at deanonymizing the account.[153]

Two days later, on February 7, 2019, the Bitcoin address, which was still active, showed that the most recent donation was that day, for $147.90. The total as of this date was $2,276.40.

On February 10, 2019, Al-Qassam Brigades reposted its poster calling for donations in Bitcoin; at that time, the bitcoin address showed $2,305 in received funds. As of February 17, 2019, it showed a total of $2,755.[154]

On February 13, 2019, it was reported that a civil rights group had threatened the leading cryptocurrency exchange Coinbase with a lawsuit if it did not put a stop to the Hamas fundraising on its platform. The group wrote: "It has recently come to our attention that the notorious Palestinian terrorist group Hamas currently maintains an account with Coinbase, Inc. through which it is accepting donations. Therefore, I am writing to notify Coinbase that knowingly providing material support or resources to Hamas is a violation of U.S. federal criminal law, and to demand that Coinbase immediately terminate any and all accounts and services provided to Hamas." Additionally, Coinbase may be in breach of its own Terms of Service; the TOS warns its users against using the service for a promotion involving violent crimes.[155] Subsequently, Coinbase shut down the Hamas fundraising campaign.[156]

Al-Qassam Brigades Release Video Of How To Donate To It Securely In Bitcoin

On March 24, 2019, the Al-Qassam Brigades released a short video describing secure methods for sending it donations in bitcoin. The clip was released as part of a social media campaign which includes posters, clips and calls from religious figures, and urges supporters to donate in cryptocurrency to the organization. The video, in Arabic with English subtitles, gave the following guidelines: The donor should first visit the donation link, fund.alqassam. net, where a QR code and wallet address (to which the donation will be sent) will appear. The donor is then instructed to copy the wallet address or scan the QR code with their cellular phone. Each user receives a unique address which is not to be shared with anyone. The donation can then be sent in one of the following methods: Method #1: The donor should ask any money exchange office to deposit the amount in the wallet address received from

The Coming Storm: Terrorists Using Cryptocurrency

the Qassam website. The donor must be sure not to mention who owns the address. Method #2: If the donor has a secure wallet, they are to transfer the money directly to the Qassam wallet. A secure Bitcoin wallet is created by using one of the trusted sites, as well as by using a public device so that the wallet is not linked to the donor's IP address. Method #3: The donor should create a new account on one of the trading platforms and deposit the donation amount into their platform account using one's credit card. Then, they transfer the amount from the platform account to their secure wallet, from where it is transferred to the Qassam wallet. If they wish to donate again, they may either transfer the money to the same wallet address previously used or receive a new address from the Qassam website, at least 24 hours after the first address was obtained. If the Qassam website is blocked in the donor's country, they should simply change one's IP address. The clip concludes with a message in Arabic to donors: "Dear donors, reserve a portion in the imminent victory and do not let your donation, no matter how small it may be, be considered disgraceful." The English subtitles read differently: "Our generous donor, your generous donation no matter how little it is, reflects that you stand by the Palestinian people."[157]

On June 8, 2019, on Telegram, Al-Qassam shared the video and posted links in other languages – Arabic, English, French, Malay, Indonesian, Russian, and Turkish. It also linked to a funding page with Quran 9:41 in English: "Go forth, whether light or heavy, and strive with your wealth and your lives in the cause of Allah. That is better for you, if you only knew." It continued: "The Israeli occupation is fighting the resistance by trying hard to block any support for it, but the resistance's friends around the world are fighting back against such Israeli attempts and seeking all possible ways to support the resistance. Your generous donation, no matter how little it is, reflects your stand by Palestinian people. We call upon friends of the Palestinians and supporters of our just cause to financially back the resistance via Bitcoin on the following wallet address." The funding page included a captcha.

The Coming Storm: Terrorists Using Cryptocurrency

كتائب الشهيد عزّ الدين القسّام

EZZEDEEN ALQASSAM BRIGADES

Chapter (9) sūrat l-tawbah (The Repentance) ❨Go forth, whether light or heavy, and strive with your wealth and your lives in the cause of Allah. That is better for you, if you only knew❩

كيف تدعم المقاومة
بعملة البتكوين

"The Israeli occupation is fighting the resistance by trying hard to block any support for it, but the resistance's friends around the world are fighting back against such Israeli attempts and seeking all possible ways to support the resistance.

Your generous donation, no matter how little it is, reflects your stand by Palestinian people

We call upon all friends of the Palestinians and supporters of our just cause to financially back the resistance via Bitcoin on the following wallet address".

validate   ☐ I'm not a robot   reCAPTCHA
Privacy - Terms

Provided by
MEMRI JTTM

Stalinsky ■ M E M R I ميمري

Below are two posters published by the Al-Qassam Brigades requesting financial support under the slogan "Support the Resistance." Both depict the Bitcoin insignia beside a defiantly raised fist, the hashtag "Support the Resistance," a QR code, and the address of the Hamas website for using Bitcoin: alqassam.net or fund.alqassam.net.

The Coming Storm: Terrorists Using Cryptocurrency

As part of a general campaign to support the "Resistance," Al-Qassam Brigades also released an animated motivational video clip on Facebook under the hashtag "Support the Resistance." The video features an Al-Qassam brigade fighter standing resolute on the battlefield, intrepidly facing enemy battleships, warplanes and tanks which target him with rockets and missiles. Despite being hit numerous times in the chest, the fighter remains standing with his arm defiantly raised in the symbol of the Resistance. The clip concludes with the following words: "We will never bend. Nothing will defeat us."[158]

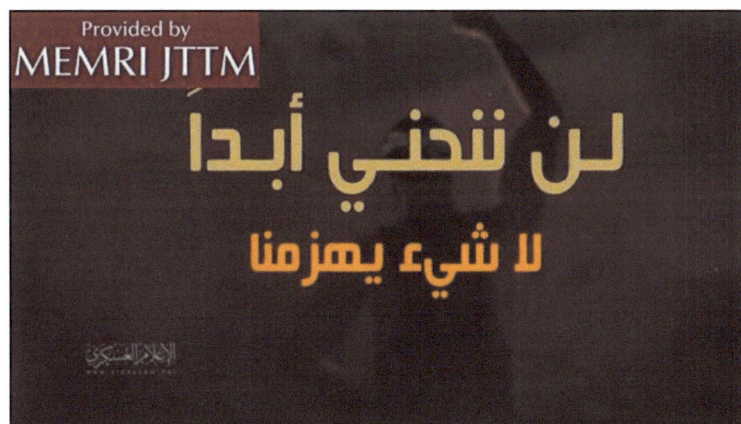

In Video, Syria-Based Jihadi Preacher Al-Muhaysini Urges Support For Hamas Via Bitcoin Donations – And Includes Al-Qassam Brigades Video Of How To Send Secure Donations Via Bitcoin

On March 26, 2019, Syria-based jihadi preacher 'Abdallah Al-Muhaysini released a two-part video on his Telegram channel in which he passionately calls upon Muslims to support the "Islamic Resistance in Palestine" (i.e. Hamas) by donating funds using bitcoin. He states: Al-Muhaysini states: "Your brothers in Gaza, your brothers the mujahideen in the Islamic resistance, have launched a campaign today to support the Resistance and its steadfastness, via what is called 'bitcoin currency.' This currency is a good, secure method by which to deliver money to mujahideen, so that you [too], Oh businessman, can engage in jihad for the sake of Allah... Oh Muslim ummah, if you cannot donate via bitcoin and are not capable of financial donations, the least you can do is spread this blessed campaign to support the steadfastness of the Resistance... Don't be parsimonious in this regard. Be a donor! May Allah be with you, our brothers in Gaza..." The second part of the video features the clip distributed by Al-Qassam Brigades on Telegram with exact instructions as to how to donate using bitcoin (see above).

In a post accompanying the video, Al-Muhaysini includes the following disclaimer: "This video is not a fatwa permitting the use of bitcoin currency. I am still hesitant about its ruling, just as our great Sheikhs are. However, we made an exception here to use it in order to overcome the security risks of donating to jihad and the like... This is the ruling of our Sheikh, the erudite Al-Didu, may Allah preserve him."

It is significant that Al-Muhaysini, a Salafi-jihadi, is encouraging support for Hamas, which is generally considered by Salafi-jihadis to lack commitment to Islamic principles.[159]

'Abdallah Al-Muhaysini passionately urging support for Hamas using bitcoin.

July 2019: Al-Muhaysini Posts Video Detailing Spending Of Nearly $200,000 Collected Via "Multiple Secure Methods" In Fundraising Campaign

On July 11, 2019, Abdullah Al-Muhaysini posted a video detailing the spending of $180,000 collected via "multiple secure methods" during a fundraising campaign that ran May 20-June 16, 2019. The video lists rebel military factions that benefited from the campaign, such as the National Liberation Front (NLF), Jaysh Al-Izza, and the Free Syrian Army.

June 3, 2019: Hamas Military Wing Al-Qassam Brigades: We'll Send "Hundreds Of Millions" Of Text Messages In Multiple Languages Explaining How To Send Funds Via Bitcoin

On June 3, 2019, Al-Qassam Brigades spokesman Abu Ubaida announced that the group would send out "hundreds of millions" of text messages to millions of people across the Arab and Muslim world, as well as to the "free people of the world," explaining how to send funds to "the resistance and jihad."

A poster on the Brigades Telegram channel said that the messages would be "accurate and trusted, aimed at simplifying the process of sending money to support resistance and jihad," and in multiple languages; they would call on the recipients to share and promote them.

The poster also expressed the Brigades' gratitude to the people of the ummah "who have been and are responding to our call to support the resistance with digital currency in our open battle against the number one enemy of the ummah."

Using the hashtag "support the resistance," the group shared multiple links to pages on its website along with a previously released video describing secure methods for sending donations to the organization using Bitcoin in languages including English, French, Russian, Turkish, Indonesian, and Malaysian.[160]

New Jersey Man Arrested And Charged With Attempts To Provide Material Support To Hamas Donated In Bitcoin Via Al-Qassam Website

One supporter of the Hamas Al-Qassam fundraising campaign was an American from New Jersey man, who was arrested May 22, 2019 and charged with attempting to provide material support to Hamas, making false statements, and making a threat against Israel supporters. He had donated in bitcoin and by MoneyGram to Al-Qassam.[161]

The Coming Storm: Terrorists Using Cryptocurrency

Authorities believed that Jonathan Xie, 20, of Basking Ridge, NJ, had intended to attack a May 31 pro-Israel event in New York's Times Square.[162] Xie had said in an Instagram Live video: "I'm gonna go to the [expletive] pro-Israel march and I'm going to shoot everybody." In subsequent Instagram posts, Xie had stated, "I want to shoot the pro-israel demonstrators... you can get a gun and shoot your way through or use a vehicle and ram people... all you need is a gun or vehicle to go on a rampage... I do not care if security forces come after me, they will have to put a bullet in my head to stop me."[163]

Jonathan Xie.

Xie had said that he would go to Gaza and join Hamas "if I could find a way." He also stated online, "Donald Trump, he should be hung from the gallows!" and, after visiting Trump Tower in New York in April, wrote on his Instagram account that he wanted to bomb it. He added, "I forgot to visit the israeli embassy in NYC... i want to bomb this place along with trump tower."[164]

In December 2018, Xie had sent $100 to an individual in Gaza he believed to be a member of the Al-Qassam Brigades. In April 2019, he sent a link to a website for the Al-Qassam Brigades to an individual with whom he was interacting online, who was an FBI employee operating undercover. Xie described the website as a "Hamas" website and then sent screenshots of the website to the undercover employee and demonstrated how to use a new feature on the website that allows donations to be sent via bitcoin. He also joked that using bitcoin was "too confusing lol." He later used the feature himself to donate.[165] Xie also repeatedly declared that he did not "give a shit" if he was arrested or killed.[166]

U.S. Attorney Craig Carpenito said, "Homegrown violent extremists like Xie are a serious threat to national security. The actions that he took and planned to take made that threat both clear and present."

The following email exchange is taken from the indictment.[167] On November 16, 2018, Xie wrote to the Al-Qassam Brigades English-language contact email asking whether non-Arabs could join and how else he could support the group. The criminal complaint does not make clear whether he received a response to this initial email, but on December 16, 2018, he wrote again to the same address asking how to donate: "Salam alaikum [Peace be unto you]! I saw at the bottom of your website that you can make a donation to the Al-Qassam Brigades by contacting this email address. As a Muslim living in the US, I would like to support the Palestinian resistance as much as possible. Thanks for taking your time to read and hopefully there's a way I can donate! "

December 16, 2018

Provided by
MEMRI JTTM

Subject: How to make a donation
From: Jon X <j******@gmail.com>
Date: 12/16/2018, 9:42 PM
To: fund@alqassam.**

Salam alaikum [Peace be unto you]! I saw at the bottom of your website that you can make a donation to the Al-Qassam Brigades by contacting this email address. As a Muslim living in the US, I would like to support the Palestinian resistance as much as possible. Thanks for taking your time to read and hopefully there's a way I can donate!

A reply came the next day, thanking Xie for supporting the "Palestinian resistance" and asking how much he intended to donate: "Waalikum As-Salam [And unto you peace] our dear brother, We hope that you are doing well Insha'Allah [Allah willing], and thanks for your support to Palestinian resistance. Regarding the donations; would you tell us the amount you intend to send so that we can send you the proper way to send it. May Allah Bless You." Xie replied: "I plan on donating $100(USD), if there are any fees necessary to send the money, I do not mind paying for those as well. May Allah Bless You as well. Ameen."

December 17, 2018

On Mon, Dec 17, 2018 at 3:31 PM alqassam fund <fund@alqassam.**> wrote:

Waalikum As-Salam [And unto you peace] our dear brother,
We hope that you are doing well Insha'Allah [Allah willing], and thanks for your support to Palestinian resistance.
Regarding the donations; would you tell us the amount you intend to send so that we can send you the proper way to send it.
May Allah Bless You

Subject: Re: How to make a donation
From: Jon X <j******@gmail.com>
Date: 12/17/2018, 7:10 PM
To: alqassam fund <fund@alqassam.**>

Provided by
MEMRI JTTM

I plan on donating $100 (USD), if there are any fees necessary to send the money, I do not mind paying for those as well.
May Allah Bless You as well. Ameen

The Coming Storm: Terrorists Using Cryptocurrency

The next email informed Xie that he could donate to the Al-Qassam Brigades "through one of the following programs: express money – money gram – western union."

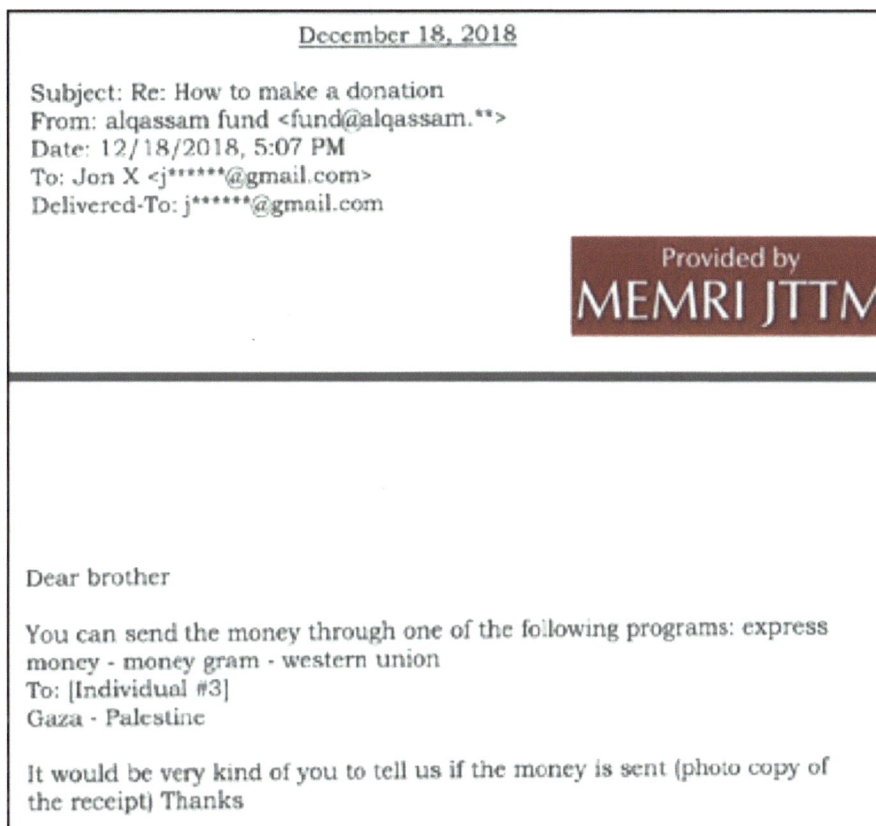

December 18, 2018

Subject: Re: How to make a donation
From: alqassam fund <fund@alqassam.**>
Date: 12/18/2018, 5:07 PM
To: Jon X <j******@gmail.com>
Delivered-To: j******@gmail.com

Dear brother

You can send the money through one of the following programs: express
money - money gram - western union
To: [Individual #3]
Gaza - Palestine

It would be very kind of you to tell us if the money is sent (photo copy of
the receipt) Thanks

On December 20, 2018, Xie replied that he was ready to make the donation via Money-Gram, but was unsure how to select the relevant territory: "Hello Brother, I am using Money Gram from my computer to send you the funds. Attached is an image of the reciever [sic] information, I just want to make sure the info I typed in is correct before I send it. In the 'Receiver country' box, I cannot find a, selection specifically for Gaza, only 'Palestinian Territory, Occupied', does this still work?" He also included an image from the MoneyGram website.

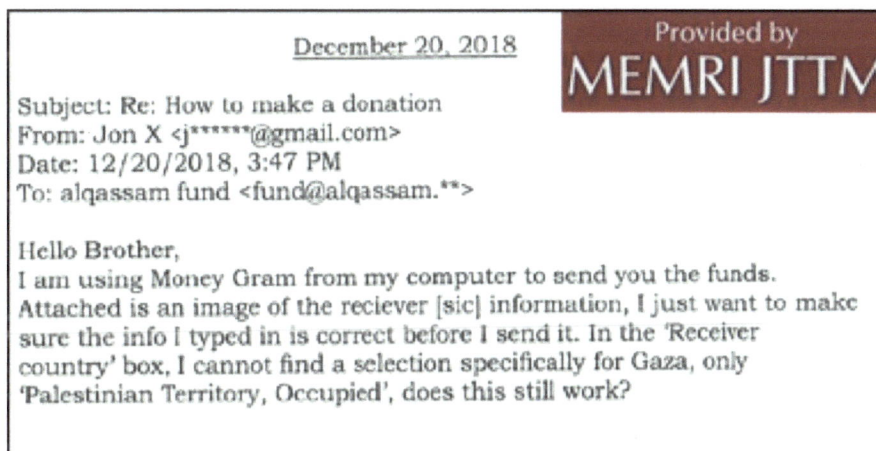

December 20, 2018

Subject: Re: How to make a donation
From: Jon X <j******@gmail.com>
Date: 12/20/2018, 3:47 PM
To: alqassam fund <fund@alqassam.**>

Hello Brother,
I am using Money Gram from my computer to send you the funds.
Attached is an image of the reciever [sic] information, I just want to make
sure the info I typed in is correct before I send it. In the 'Receiver
country' box, I cannot find a selection specifically for Gaza, only
'Palestinian Territory, Occupied', does this still work?

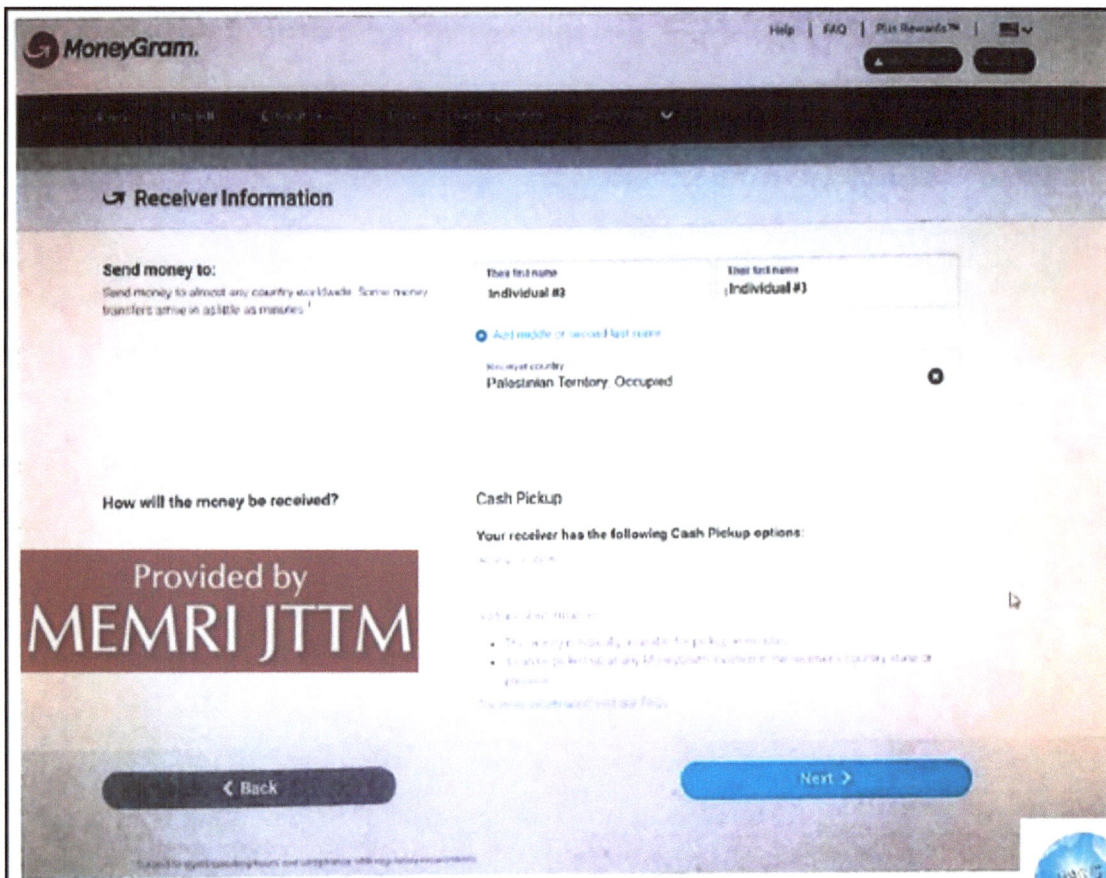

The respondent assured Xie that Al-Qassam Brigades would be able to receive the donation if he selected "Palestinian Territory, Occupied."

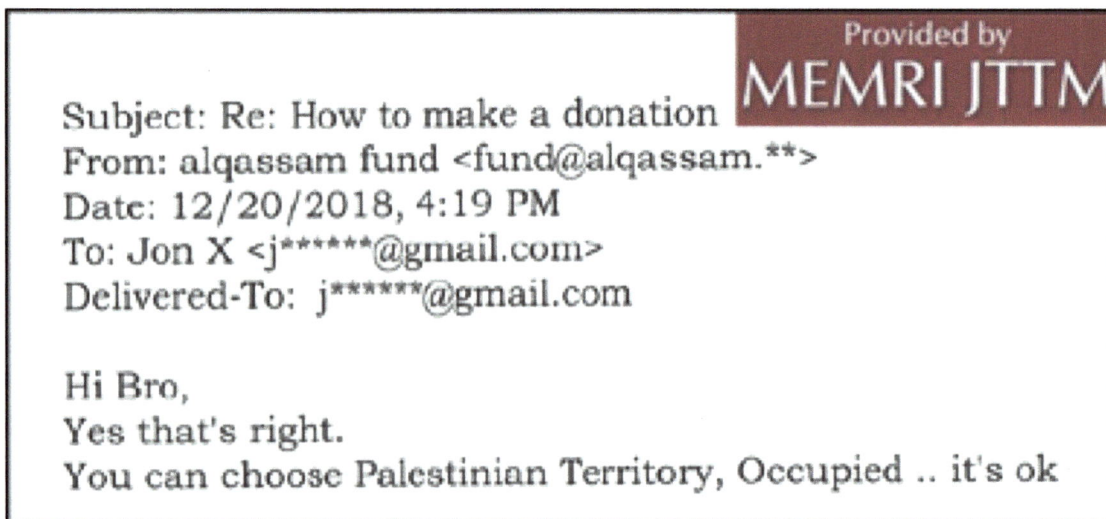

Subject: Re: How to make a donation
From: alqassam fund <fund@alqassam.**>
Date: 12/20/2018, 4:19 PM
To: Jon X <j******@gmail.com>
Delivered-To: j******@gmail.com

Hi Bro,
Yes that's right.
You can choose Palestinian Territory, Occupied .. it's ok

On December 20, 2018, Xie successfully completed the transaction and sent images of the transaction details, writing: "Okay, the transaction went through. In the 2nd picture, the

The Coming Storm: Terrorists Using Cryptocurrency

Reference Number is 37337500 when you go to recieve [sic] the cash pickup. I hope everything works out fine inshallah [Allah willing]. Let me know if you were able to recieve [sic] it, brother."

Subject: Re: How to make a donation
From: Jon X <j******@gmail.com>
Sent: Thursday, December 20, 2018 10:21 PM
To: alqassam fund

Okay, the transaction went through. In the 2nd picture, the Reference Number is 37337500 when you go to recieve [sic] the cash pickup. I hope everything works out fine inshallah [Allah willing]. Let me know if you were able to recieve [sic] it, brother.

Receiver Information	
Name:	Individual #3
Expected Destination:	Palestine
Receive Option:	Cash Pickup
Reference Number:	37337500 (Used by MoneyGram for tracking)
Date Available in Expected Destination:	20-Dec-18 (May be available sooner)
Transfer Amount:	100.00 USD
Transfer Fees:	+ 12.00 USD
Transfer Taxes:	0.00 USD
Total Cost:	112.00 USD

Two days later, on December 22, 2018, Xie received an email confirming that the donation had been received and thanking him: "Dear brother, We write to tell you that we received your donation May Allah bless with his bounty of barakah [blessings] for yourself and your family"

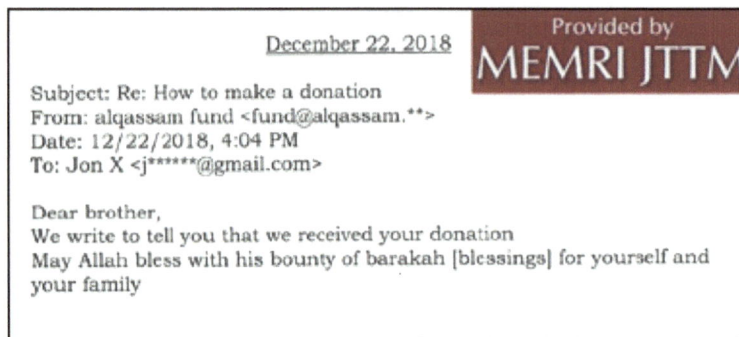

Xie then thanked the respondent, and expressed wishes that "the Palestinian resistance will be able to overthrow the apartheid Israeli regime" along with a photo of a Hamas flag.

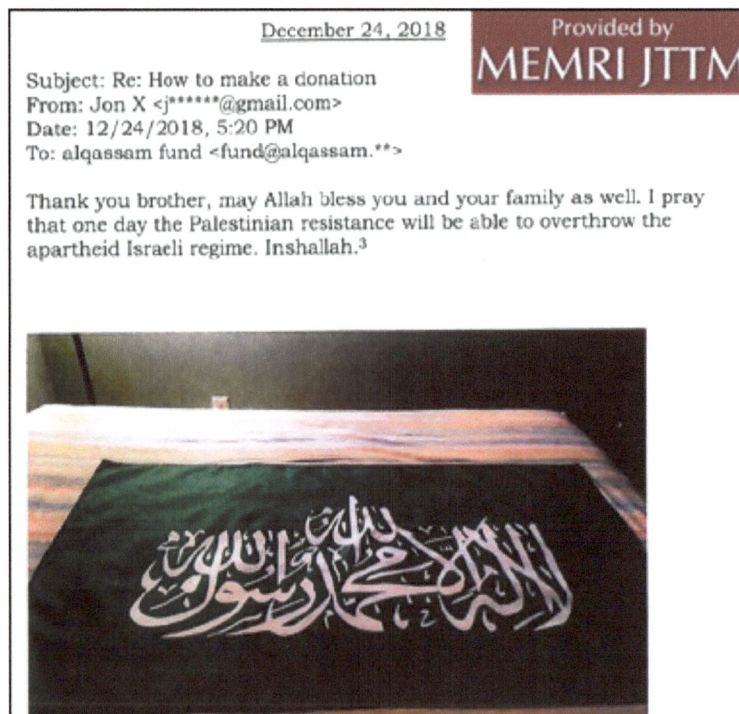

Xie also made a number of statements detailing his support for Hamas via Instagram, according to the criminal complaint. On January 19, 2019, Xie wrote on his Instagram account: "I'm going to donate more money to Hamas once I get my paycheck."

Also on Instagram, Xie revealed his plans to join the army: "I'm joining the Army... I [sic] said I could get arrested... If they found out about the Hamas shit... I'm joining the US Army not to fight foe Jewish internets [likely "interests"] But to learn how to kill So I can use that

The Coming Storm: Terrorists Using Cryptocurrency

knowledge I already got accepted Israel does that so they can kill innocent women and children Idk [I don't know] if I pass the training If I should do lone wolf That is why I have to learn military techniques from the Army To stop these people"

February 18, 2019

Provided by MEMRI JTTM

Xie: I'm joining the Army . . .
I [sic] said I could get arrested . . . If they found out about the Hamas shit. . .
I'm joining the US Army not to fight foe Jewish internets [believed to be a typo of "interests"]
But to learn how to kill
So I can use that knowledge
I already got accepted
Israel does that so they can kill innocent women and children
Idk [I don't know] if I pass the training
If I should do lone wolf
That is why I have to learn military techniques from the Army
To stop these people

In his April 2019 interaction with the individual who was an undercover FBI employee, Xie provided, as noted, specific instructions on how to donate to Hamas, and again confirmed his own donations. The images below include excerpts from the interaction as well as images Xie sent to the FBI employee (Individual #4). Xie explicitly mentions bitcoin in response to a question about the risk of getting caught, saying: "if you click the left button, it brings you to their donation page where you can donate via bitcoin"

April 14, 2019

Provided by MEMRI JTTM

Individual #4: Does Hamas have good media? I never see much by them.

Xie: I think it's pretty decent.

***

Xie: actually let me send you the link to al-qassam's website

Individual #4: Where can I find them?
Okay, is it safe?

Xie: Yeah I'm pretty sure its safe

Xie: I've used it a bunch of times and nothing bad has happened so far

Xie: https://alqassam.net[4]

Xie: They changed the layout of the website a bit

Xie: If you click the left button, it brings you to their donation page where you can donate via bitcoin

Xie: and the right button is for their main website

Individual #4: Is it actually Hamas site?
Like run by them?

Xie: I'm like 95%+ sure

Individual #4: I wonder if donations actually go to the mujahideen[5] or like the political stuff

Xie: They also have another website, but that's for the political stuff

When Individual #4 expressed confusion over the process, Xie sent pictures with instructions on how to donate: "Oh! I found how to translate the donation page to English... ok ill send you pics."

Assorted Jihadi Telegram Channels' Ongoing Efforts In Fundraising In Bitcoin And Other Cryptocurrencies

Pro-ISIS Channel Shares Guide, Links To Purchase Weapons On Dark Web Using Bitcoin, Urges Muslims To Attack In The West – "Make The Unbelievers Bleed Inside Their Cities"

On April 11, 2019, a pro-Islamic State (ISIS) Telegram channel posted a message in English urging Muslims in the West to buy weapons on the Dark Web and carry out terror attacks in revenge for the killing of ISIS fighters in Baghouz, Syria. It asked readers, "You wanna make kuffar bleed? You wanna make them pay for killing children and woman in #Baghouz and in every state of the #Dawlah?" The message was shared across several ISIS channels with a hashtag referencing the recent ISIS-initiated campaign of attacks labeled the "Blessed Raids Avenging Syria." It also included a tutorial from a non-jihadi technical source explaining how to use the Dark Web, bitcoin, and secure email services for purchasing weapons on the Dark Web.

A recent ISIS message on Telegram addressing the "lone wolves in the land of unbelievers" referred them to "the best armory in the Dark Web" – "Lucky 47 Luhansk Counter Kiev Partisans," or LUCKP47SHOP, that promises to deliver weapons worldwide and provides a bitcoin address. This marketplace bills itself as "a paramilitary organization... [currently fighting] against the massacre on the domestic population Luhanska" and provides photos of weapons offered for sale, including RPGs and RPG ammo, mines, grenades, and small arms. It states: "All weapons are fully functional. Our products we have stored directly in the country of the EU." The message urged readers to to buy any weapons on the site and then carry out attacks and "make the unbelievers bleed inside their cities."[168]

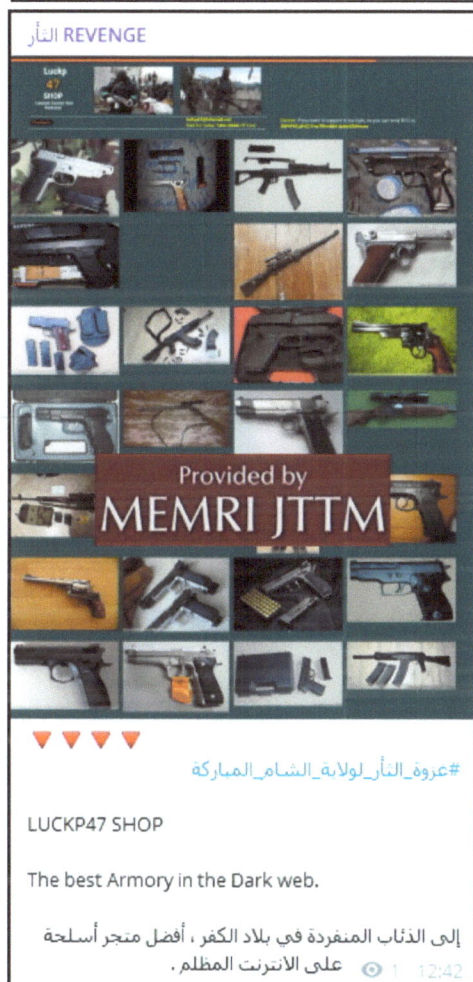

Addressing the "lone wolves in the land of unbelievers," the post refers them to "the best armory in the Dark Web."

Provided by
MEMRI JTTM

"Race One With Another To Paradise" In English Recruits, Fundraises For Jihad In Syria By Using Bitcoin, PayPal And Selling Merchandise; Gives Tips On Cybersecurity, Traveling To Syria

The English-language Telegram channel "Race One With Another To Paradise" recruits and fundraises for jihadi purposes and is operated by a jihadi fighter and media operative likely based in Syria. Posts indicate that donations to the channel's Bitcoin and PayPal accounts fund the mujahideen in Syria and their widows and children, and purchase supplies and media activity. While there have been several versions of the Telegram channel, the most recent version was opened on August 28, 2018[169] and has 254 members. The channel shares content that favors Hay'at Tahrir Al-Sham (HTS) and other groups in Syria. Several messages give "tips for hijra [immigration]" and links to GoFundMe accounts for other Islamic projects. The person operating the channel says that he has spent time in Europe. Other accounts also share its posts.

The Coming Storm: Terrorists Using Cryptocurrency

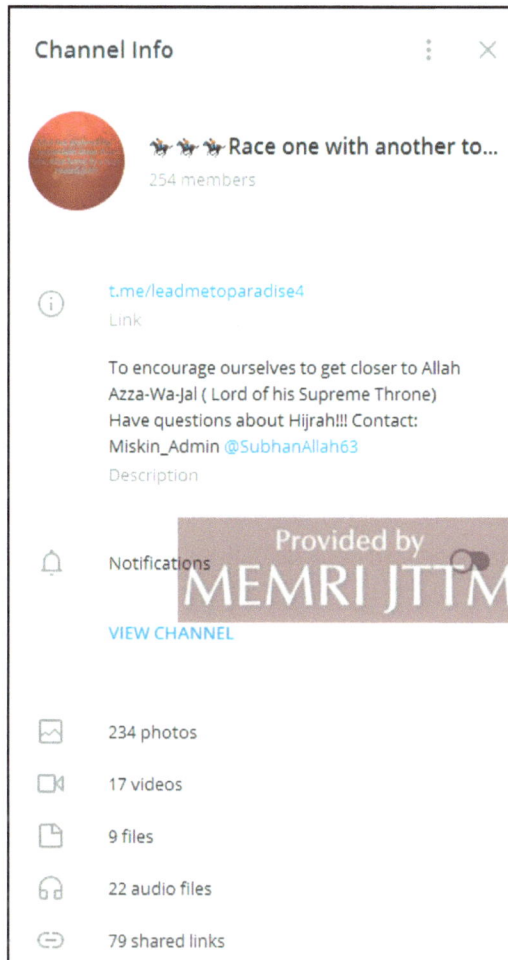

Channel Info

🐎🐎🐎Race one with another to...

254 members

t.me/leadmetoparadise4
Link

To encourage ourselves to get closer to Allah
Azza-Wa-Jal (Lord of his Supreme Throne)
Have questions about Hijrah!!! Contact:
Miskin_Admin @SubhanAllah63
Description

Notifications

VIEW CHANNEL

234 photos

17 videos

9 files

22 audio files

79 shared links

Channel info page for Race One With One Another To Paradise.

The channel regularly calls for Muslims to join the jihad in Syria. It offers to help those interested in traveling to Syria to fight, and encourages supporters to donate money to aid the mujahideen, saying that donations can be made via PayPal and Bitcoin accounts. The contact information provided is a Telegram account, @SubhanAllah63.

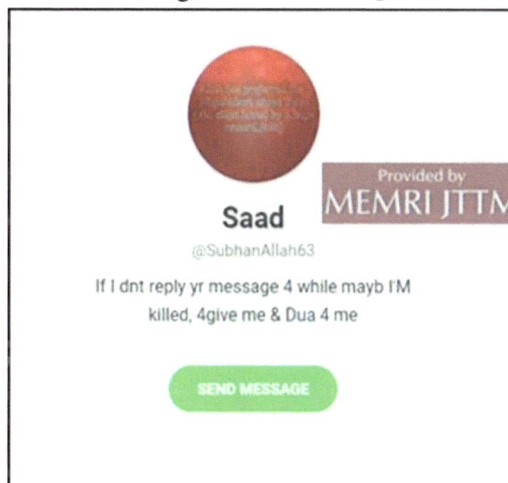

Saad

@SubhanAllah63

If I dnt reply yr message 4 while mayb I'M
killed, 4give me & Dua 4 me

SEND MESSAGE

Fundraising Using Bitcoin, PayPal, And Selling Merchandise

On September 1, 2018, the channel posted a message calling for donations in advance of a possible battle for Idlib: "If this attack starts, it will be a big one and history will be made. I request from all of you not to sit on the sidelines. Do participate with the brothers. Unbelievers, polytheists and hypocrites are united while we are divided. We have limited resources, money. If anyone wishes to participate in this jihad for Allah's sake by giving as much as they can, please contact @SubhanAllah63. Your identity will not be disclosed."

On November 29, the channel wrote: "Admin Note: Tawheed [monotheism] T Shirts for Sale. One of our friends launched his new Islamic Tshirt designs. The profits from these sales are supposed to be for a good cause..." The t-shirt is being sold at teespring.com/Tawheed2.[170] According to the website, the "products are fulfilled in the U.S." and the "Campaign ID" is Tawheed2.

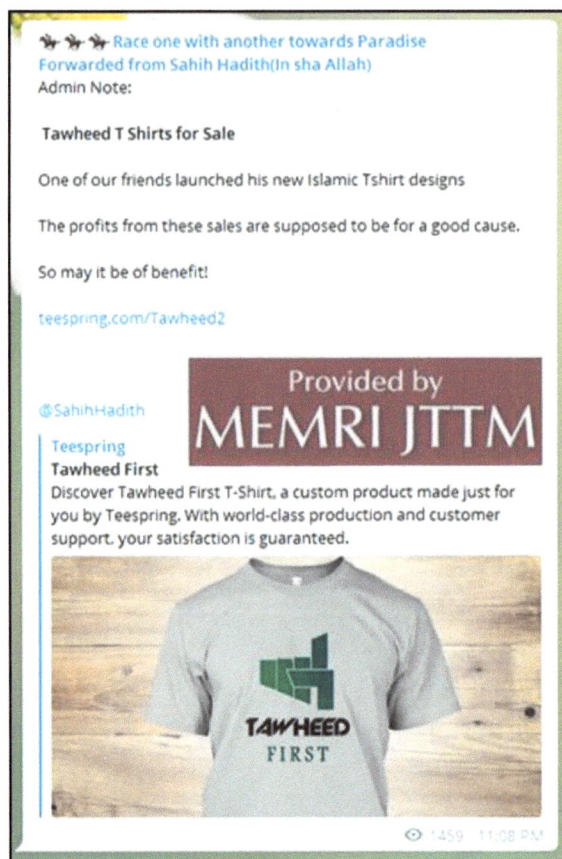

The channel posted several such messages encouraging Muslims to support jihad, writing on August 29, 2018: "We have PayPal and Bitcoin account available. Please for the sake of Allah don't just sit and do nothing."

The Coming Storm: Terrorists Using Cryptocurrency

Recruiting And Encouraging Hijra ("Immigration")

On November 17, the channel administrator posted a message saying: "If you are young, if you feel you are active, you are not among sleeping ummah, & you like to do something amazing for ummah of Muhammad (saw), you like to join the ranks of Mujahideens then feel free to contact here: @SubhanAllah63. Maybe we will have something for you to do for Mujahideens of Shaam, not ISIS, we don't work with isis."

On October 7, 2018, the channel wrote: "Dear brothers and sisters in Islam. Praises to Allah. #wish to make hijrah to Syria!!! #wish to sponsor an orphan from Syria!!! #wish to sponsor a widow (husband killed in battle for Sake of Allah)!!! #wish to sponsor a mujahid (freedom-fighter)!!! Please DM/PM here: @SubhanAllah63. Need proof for your investment? No problem, we will show you."

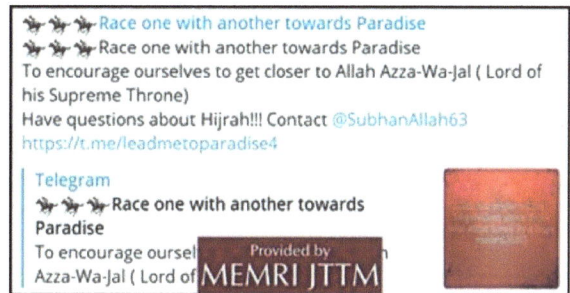

A post promotes channel and offers to answer questions about traveling to Syria.

On October 23, the channel wrote: "Dear brothers and sisters in Islam...# wish to make hijrah to Shaam !!! # wish to sponsor a orphan from Shaam !!! # wish to sponsor a widow (husband killed in battle for Sake of Allah Azza-Wa-Jal) !!! # wish to sponsor a mujahid (freedom-fighter) !!! Please DM/PM here: @SubhanAllah63."

Stalinsky ■ MEMRI ■ ميمري

On November 6, the channel shared tips on traveling to Syria, including: "Take your time, make your plan (how to leave country without even showing any sign to authority), take advices from brothers who are already inside Shaam, don't be hasty, don't share your plan with anyone except people you think will never disclose your plan with their friends, prepare some amount of money that you will need for your travel...pray a lot to Allah Azza-Wa-Jal to guide you safely to right Jamah [congregation]." The administrator says it took them about 2 1/2 to 3 years to "make up everything. Now here I'm where I wanted to be."

Tips On Cybersecurity And Remaining Anonymous Online

The channel administrator shares tips telling readers how to remain anonymous online, adding that they have "a computer science background." On August 28, 2018, the channel posted: "Few tips for those who work online for the sake of Allah [...] if you want to save [i.e. hide] your identity online from anyone could be government intelligence, Google, Telegram, etc... Don't be scared. There are ways to save your online identity."

On December 19, the channel wrote: "Warning...!!! When you come virtual life like Telegram/wtsapp/etc don't share your personal info to anyone."

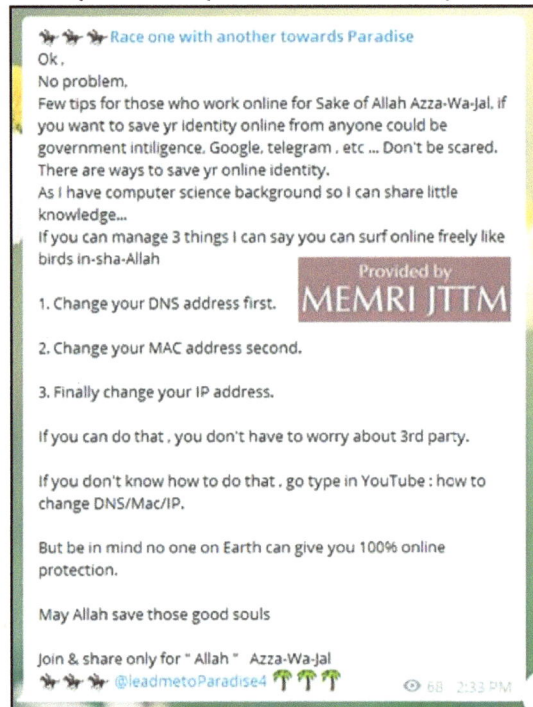

The Coming Storm: Terrorists Using Cryptocurrency

Encouraging Jihad And Denouncing Kuffar ("Unbelievers")

On August 28, 2018, the channel wrote that those who act to "make the unbelievers angry/irritated/uneasy" will gain spiritual reward.

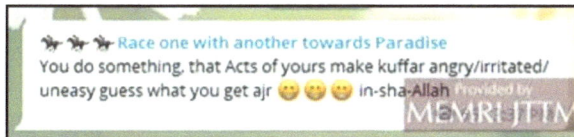

The channel wrote on October 19: "Dear brothers and sisters in Islam...We will strongly recommend you engage yourself to such activities that will give you elite rank in hereafter...Help the religion of Allah Azza-Wa-Jal. Either you help or not this religion gonna victorious without any doubt. But the point is are your elite place secure in *Jannah* [Paradise]!!!"

The channel posted an image on November 13 that reads: "Allah has preferred the Muja-hideen over those who remain [behind] with a great reward. Degrees from Him and forgiveness and mercy. And Allah is ever Forgiving and Merciful. The Messenger of Allah was asked, 'Who is the best of people?' He said, 'A believer who is making Jihad with his life and his wealth in the cause of Allah.'"

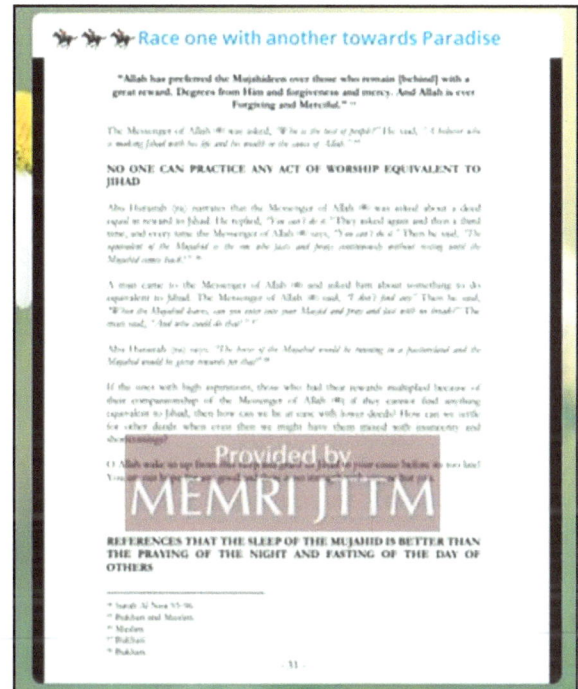

On November 29, the channel posted a message showing a hypothetical conversation in which a person says they "want to fight against enemy of Islam" but then says they are afraid to send money to the mujahideen because they are "scared of enemy of Islam, they might arrest me."

On November 4, the Telegram channel wrote: "...Brothers & sisters in Europe & developed countries, this luxurious life will soon finish, it's like Allah is giving you choice. You can have a luxurious life is you want & you can have a struggling life if you want but the reward is not the same. Work according to your rank in Jannah."

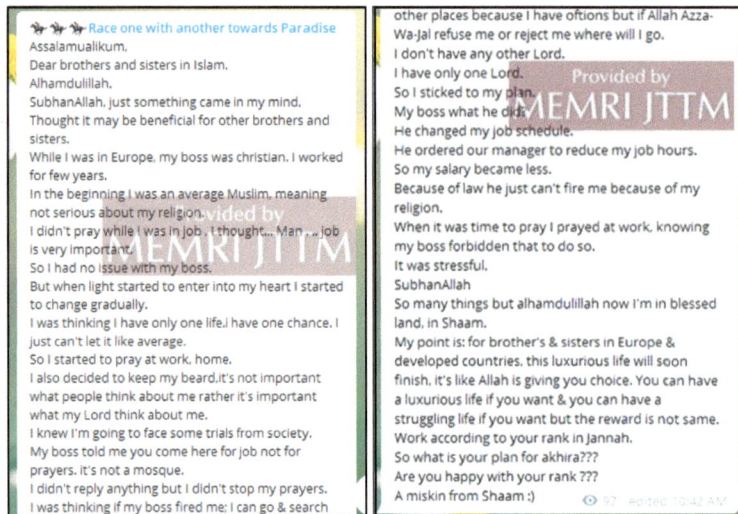

Promoting Other Groups

On October 23, the channel posted a link to a YouTube video called "Things That Allah Likes" along with a link to a GoFundMe page.[171]

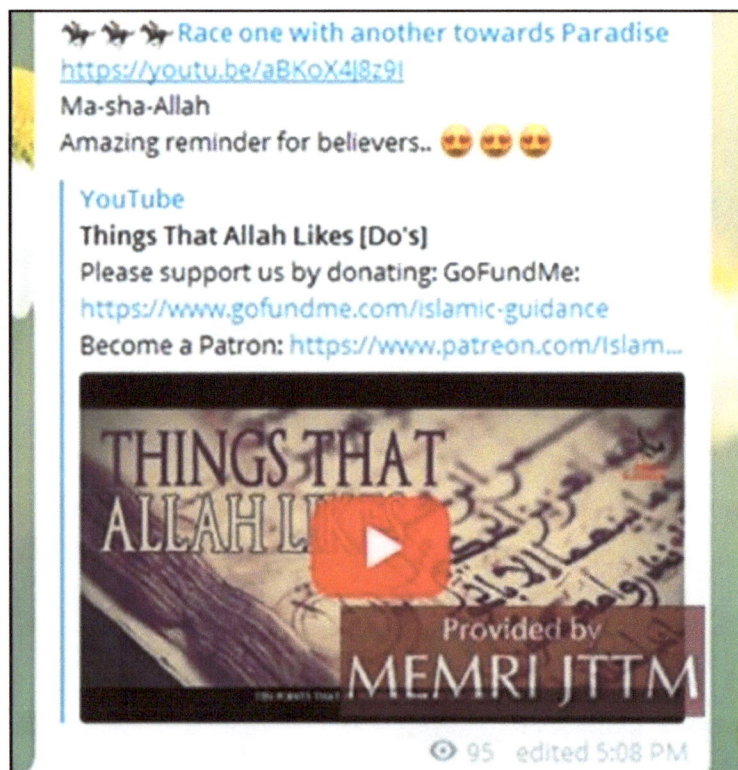

The Coming Storm: Terrorists Using Cryptocurrency

The GoFundMe page is for a YouTube channel created in 2008 focused on Islam called Islamic Guidance. The GoFundMe page has raised 13,948 euros in the last two years. The platform is run by one Shek Shamim Ahmed and is based in England. Islamic Guidance has 440,000 subscribers and 362 people have donated to help in the making of their videos.

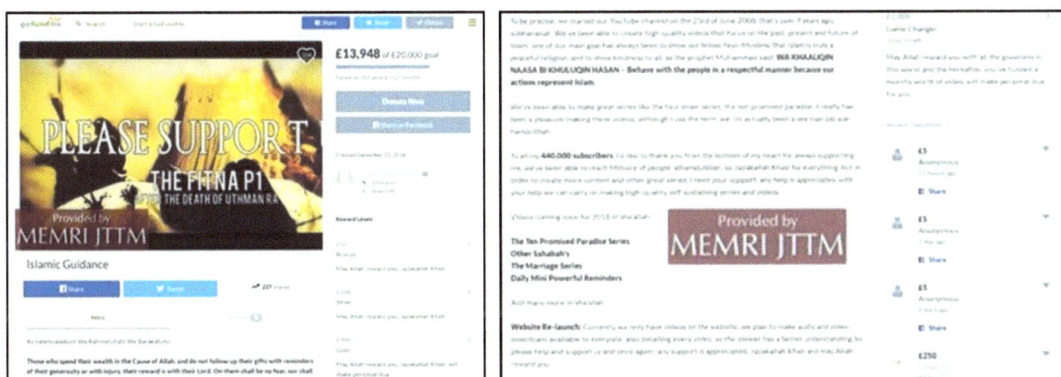

On November 4, the channel shared a link to another GoFundMe page.[172]

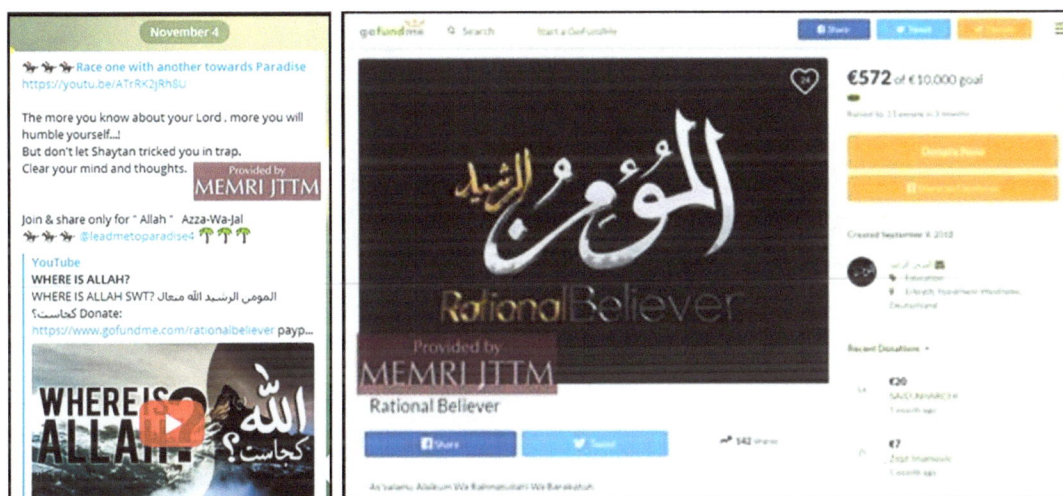

The GoFundMe page was for Rational Believer, a YouTube channel based in Germany that uploads videos on Islam and has 62,000 subscribers. The channel has raised 572 euros from 23 donors in the past three months. The donations from this GoFundMe page are used to "upgrade our current equipment and to setup a budget (for our production costs) to provide you higher quality contents and to upload more frequently."

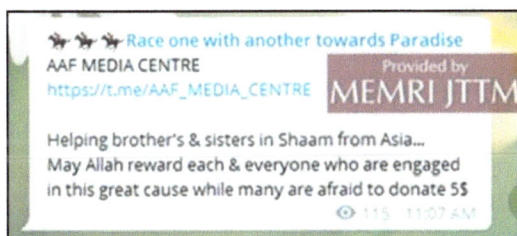

On October 14, the channel posted a link to the AAF MEDIA CENTRE Telegram channel and wrote: "Helping brothers & sisters in Shaam from Asia...May Allah reward each & everyone who are engaged in this great cause while many are afraid to donate 5$." The link goes to a Telegram channel belonging to the Abu Ahmed Foundation, an

Indonesian fundraising group that raises money to buy food, weapons, clothing, and other materials for jihadis.[173]

On October 23, the channel posted a link to a YouTube video[174] of an On The Ground News interview with a Chechen jihadi commander. The Telegram channel wrote of the interview: "He said some strong and powerful words superbly. You want victory of ummah while those who are in ground fighting & sacrificing them but when they return home they have no Food to eat, no house to rest, no weapons to terrorize the *kuffar* [unbelievers]. This is not shame for us??? You have always wealth to fulfill your needs...but when it comes to fi-sabilillah you drop 5$ & you think you have done some amazing job by donating big amount of 5$."

Channel's Instructions On Using Bitcoin, Other Cryptocurrencies Disseminated By Others

In June 2019, the Life_of_Mujahideen_in_Shaam channel shared Race One With Another Towards Paradise posts with information on how to buy bit-

The Coming Storm: Terrorists Using Cryptocurrency

coin "easily & securely without getting fraude by scammers" and also how to use. It describes how to install the Blockchain app, activating two-factor authentication (2FA) and refers readers to a YouTube video for more information, and also gives information on opening a Paxful account for buying bitcoin.

The "channel info" for Life_of_Mujahideen_in_Shaam" states: "Send some bullets here by sending some money for Mujahideen in Shaam: @forahlealsunnahjamah"

In late July 2019, the channel wrote: "We are collecting donations for needy in Shaam from our generous brother's & sister's who are outside Syria" and inviting followers to donate for "poor family or orphans or widows or mujahideen" via "Western Union or Paypal or Bitcoin (most secure and safe in-sha-Allah)."

Life_of_Mujahideen_in_Shaam
1,103 members

Pinned message
Save this Telegram I'd : @forahlealsunnahjamah Reason: if you see our ⊙

Life_of_Mujahideen_in_Shaam

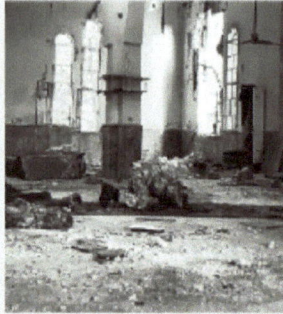

Life_of_Mujahideen_in_Shaam

From.............................

To.................................

Jaza-kallahu-khaer for your donations :)

We are collecting donations for needy in Shaam from our generous brother's & sisters who are outside Syria.

If you wish to donate for:

Poor family
Or
Orphans
Or
Widows
Or
Mujahideen

Then kindly drop your message here: @forahlealsunnahjamah

Have no doubt about us , if you have any then check our previous contents or message us for evidence/proof.

You can donate via: Western Union or PayPal or Bitcoin (most secure and safe in-sha-Allah)

"Who is it that would loan Allah a goodly loan so He may multiply it for him
many times over? And it is Allah who withholds and grants in abundance, and to
Him you will be returned."

[Surah al Baqarah 245]

⚔ @lifeofmujahideensinshaam ⚔   👁 452   edited 8:44 AM

The Coming Storm: Terrorists Using Cryptocurrency

"The Merciful Hands" Fundraises Using Bitcoin, Other Methods For Jihadis In Syria, In English, German, Turkish, And Arabic

A jihadi Telegram channel called The Merciful Hands says it is "giving aid to those in need."[175] The channel lists a Bitcoin address and a WhatsApp number with a Turkish country code (+90 537 822 0931) for donations to fighters in Syria, but these numbers can be faked. The Merciful Hands solicits donations in English, German, Turkish, and Arabic. There are also posts in Russian. On March 28, 2018, The Merciful Hands was promoted by the pro-Al-Qaeda Telegram channel Ummah Chat Group. The Merciful Hands was created on March 20, 2018, and as of this writing has 1,212 members. There are photos on the channel of donors holding up the US dollars that they donate, as well as photos of the people in Syria who receive the donations.[176]

The channel's first post, which has since been deleted, described on March 20 the financial situation of four men in Syria. The post read: "I am writing this message today because I have seen the financial situation first hand of many *muhjireen* [immigrants, i.e., foreign fighters] and *ansar* [natives]. The situation is deteriorating everyday as you know the *mazoot* [fuel oil] 650 a liter a *jarrah* [container] gas is nearly 15,000 lyra to fill your *cazan* [tank] you have to pay for the tractor to come to your house so this increases the price of the water.

"So what I am proposing is to try to help some brothers I know first hand first we will start off with 5 brothers and later if this can work we will open this to more brothers, the money is an *Ammana* [responsibility] and we are scared of *yamal qiyamah* [Day of Judgement] the so the money wont be misappropriated. The brothers I wont list their names for security purpose and I don›t want people know they are in need since they are such humble brothers...

"1. The first brother is a muhjir murabit [one who goes on guard duty] and goes to *ribat* [guard duty] when the call for jihad or ribat is made. This brother is married recently only 4 months he spent all his money on the *mahar* [dowry] and he needs to buy a bike to get him and his wife around. So if someone wants to donate a car, bike, or money please feel free to contact us.

"2. The second brother he is a *muhajir* [migrant i.e., foreign fighter] as well he is not married but is not part of any group because he is dealing with *sihr* [sorcery] so for him to go to the battlefield would only lower the moral of the brotherse because of the fear the *jinn* [spirit] gives him. The brother barely has 2l lyra in his pocket at times and doesn›t get help from his country or from any group.

"3. The third brother is also a mujjir and he works media only because he was injured and shot with a sniper bullet in his leg, but as he says when he is top shape he will return to the field also he is married with one child and a new baby on the way many time the brother can barely afford to pay for milk or diapers, at one time his wife mixed leban and water to try to give it to the baby and the baby ended up getting sick this is how dire their situation is..."

Asslam alak[...] akatu ya brothers and sister in islam

I am writing this message today because I have seen the financial situation first hand of many muhjireen and ansar. The situation is deteriorating everyday as you know the mazoot 650 a liter a jarrah gas is nearly 15,000 lyra to fill your cazan you have to pay for the tractor to come to your house so this increases the price of the water.

So what I am proposing is to try to help some brothers I know first hand first we will start off with 5 brothers and later if this can work we will open this to more brothers, the money is an Amman and we are scared of yamal qiyamah the so the money wont be misappropriated. The brothers I wont list there names for security purpose and I dont want people know they are in need since they are such humble brothers and just have tawwakal in Allah azza wa jal

1. The first brother is a muhjir murabit and goes to ribat when the call for jihad or ribat is made. This brother is married recently only 4 months he spent all his money on the mahar and he needs to buy a bike to get him and his wife around. So if someone wants to donate a car, bike or money please feel free to contact us

2. The second brother he is a a muhajir as well he is not married but is not part of any group because he is dealing with sihr so for him to go to the battlefield would only lower the moral of the brothers because of the fear the jinn gives him. The brother barely has 2k lyra in his pocket as times and doesn't get help from his country or from any group.

3. The third brother is also a muhjir and he works media only becuase he was injured and shot with a sniper bullet in his leg, but as he says when he is top shape he will return to the field also he is married with one child and a new baby on the way many time the brother can barely afford to pay for milk or diapers, at one time his time his wife mixed leben and water to try to give it the baby and the baby ended up getting sick this is how dire there situation is. He received a monthly ratib of 20,000 lyra a month but its not enough to take care of his family for one month so this is why we are asking for donation.

The Coming Storm: Terrorists Using Cryptocurrency

The channel posted a photo of a person holding a $100 bill and notebook that read "Sadaqah Al-Khair, Merciful Hand Syria," also on March 20.

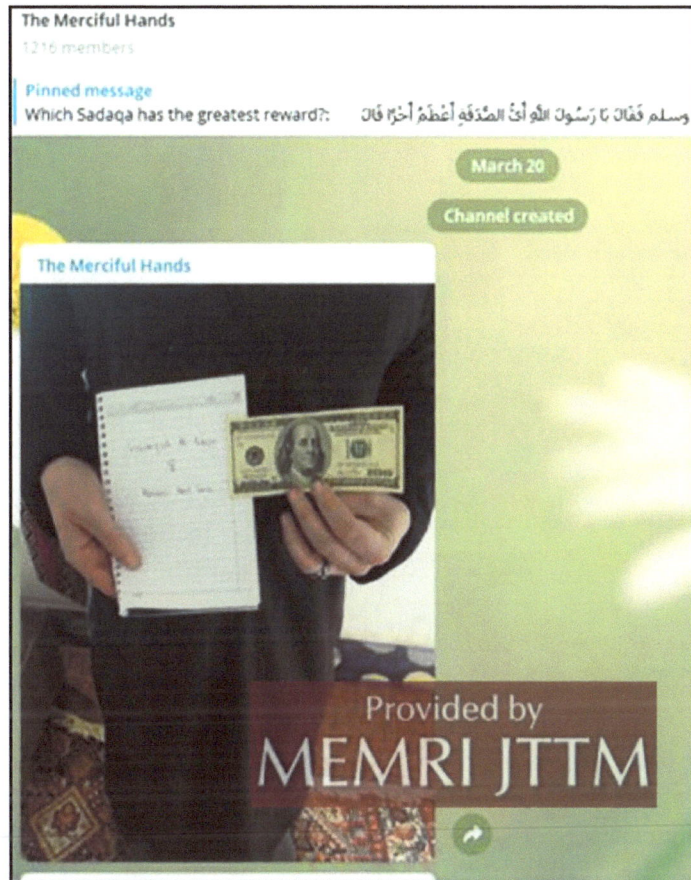

On March 21, a photo was shared showing a man holding a $50 bill with a hand-written note: "The Merciful Hands from Sham to the Muslims of South Africa, we thank you for feeding us." The photo appears to show a recipient in Syria of a donation from South Africa.

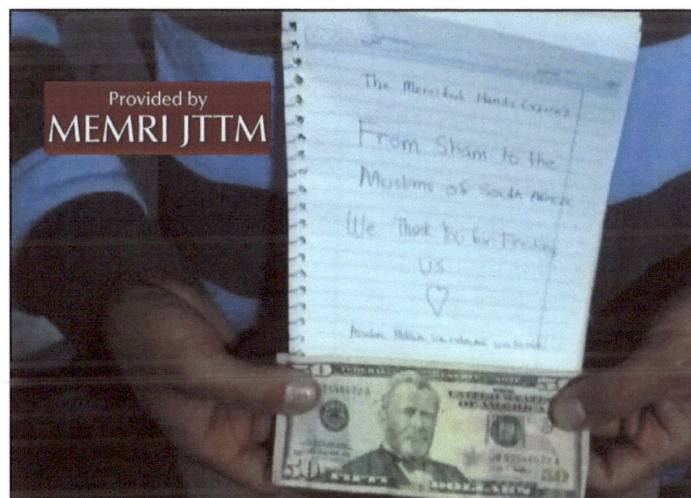

On March 23, a post about two men was shared on the channel. The post read: "A sad story of foreigners forgotten in the land of Sham [Syria].

The Merciful Hands
Assalam alakium wa rahmutalhi wa barakatu

A Sad story of foreigners forgotten in the land of sham 😔

1. A brother who is in the cause of Allah wife is sick they have a child and they are not getting any finacial support for only 4 months they got $30 the brother has sold all his equitment, his bike, his phone, pretty much everything worth of value take care of his family so please find it jn your hearts to help thjs brother. Even though he is not getting any money hes still staying in thr field.
Jazakallah Khair

2. The second brother is in a tough finacial situation hed getting kicked out of his house with no place to go he doesnt have any electricity, and only uses a lighter for light he wad recently married 3 months ago
this brother tore his shoulder snd has pain in his back from disk also nerver damage in a paet of his body, but these injurys didnt deter the brother from being in the cause.

Final note:
these two brothers are in extreme fiancial poverty going months without food boxes snd payment. So lets work together to get these brother something so they can have some assistance.

Provided by
MEMRI JTTM

"1. A brother who is in the cause of Allah wife is sick they have a child and they are not getting any financial support for only 4 months they got $30 the brother has sold all of his equipment, his bike, his phone, pretty much everything worth of value take care of his family so please find it in your hearts to help this brother. Even though he is not getting any money he's still staying in the field.

"2. The second brother is in a tough financial situation he's getting kicked out of his house with no place to go he doesn't have any electricity, and only uses a lighter for light he was recently married 3 months ago. This brother tore his shoulder and has pain in his back from disk also nerve damage in a part of his body. But these injuries didn't deter the brother from being in the cause. Final note: These two brothers are in extreme financial poverty going months without food boxes and payment. So let's work together to get

these brother something so they can have some assistance."

It appears that the person who wrote this post intentionally avoided using the word "jihad" and instead used the phrase "the cause."

Another post from March 23 read: "There is a brother who is new to the field and has no money to purchase equipment vast [sic] something for protection when going to the fronts. So the brother is looking for donations. I have written this without using specific words but you know what I mean ;) so in sha Allah if there is anyone available to help please contact us so we can help get this brother to the field. One who finances someone in the field is like the one who is in the field himself. Don't worry about donations small or big everything counts and your reward is with Allah in sha Allah." The post is signed: "Suliman Al Hindi Telegram - +90 537 822 0931," which indicates that the person who wrote the post may be Indian.

The Merciful Hands
السلام عليكم ورحمة الله وبركاته
There is a brother who is new to the field and has no money to purchase equitment example vast something for protection when going to the fronts.

So the brother is looking for donations. i have written this without using the specific words but you know what i mean :) so in sha Allah if there is anyone available to help please contact us so we can help get this brother to the field.
one who finances somone in the field is like the one who is in the field himself. Dont worry about donations small or big everything counts and your reward is with Allah in sha Allah
jazakallah khair.

Provided by
MEMRI JTTM

السلام عليكم ورحمة الله وبركاته
هناك اخ جديد في الساحة وليس لديه اي مال ليشتري العدة للحماية على الجهات وما الى ذلك فلاخ يحتاج إلى التبرعات
فقد كتبت هذا من غير استخدام الكلمات المخصصة ولكن تعلمون المقصد فان شاء الله ان كان هناك احد قادر على مساعدة الاخ نواصل معنا كي نرسله الى الجهات
ومن جاهز غاربا فقد غزى واي تبرعات صغيرة كانت او كبيرة كل خير واجركم عند الله وحزاكم الله خيرا

نواصلوا معي هنا على نلقرام او على وانس

٠٠٩٠٥٣٧٨٢٢٠٩٣١   👁 251   12:48 PM

The Coming Storm: Terrorists Using Cryptocurrency

A post in the channel from March 26 discussed two converts to Islam who immigrated to Syria to join the jihad. The post read: "Another story of two mujahirin brothers. Both brothers used to be Christians and Alhumdulilah Allah guided them to Islam and after that guided them to make *hijrah* [immigration] for jihad. Both brothers are married with two kids and one is expecting his third and as their families are *kuffar* [unbelievers] obviously they do not help them at all. Both brothers were in debt with one brother recently selling most of his house stuff to pay off his debts but now they are both broke. So if there's anyone who can help them please contact us and may Allah reward you."

The Merciful Hands
Another story of two muhajirin brothers.
Both brothers used to be Christians and Alhamdulillah Allah guided them to Islam and after that guided them to make hijrah for jihad. Both brothers are married with two kids and one is expecting his third and as their families are kuffar obviously they do not help them at all. Both brothers were in debt with one brother recently selling most of his house stuff to pay off his debts but now they are both broke. So if there's anyone who can help them please contact us and may Allah reward you.

Provided by
MEMRI JTTM

On December 6, 2018, the channel posted a new Bitcoin address 1FeocjAnzCY4BEB7pDotUAXFK9fv9774zf. The address shows no transactions.[177] It also gave the Telegram number +90 537 822 0931, the Whatsapp number +46 73 929 9885, and the Telegram admin @ thefigtree.

So we appeal to you today to donate whatever you can where it be a euro a dollar 100 dollars whatever you can your effort is appreciated and should never be looked down as it is not enough.

To donate
Telegram +90 537 822 0931
Whatsapp +46 73 929 9885
Telegram Admin @thefigtree
Bitcoin address: 1FeocjAnzCY4BEB7pDotUAXFK9fv9774zf

Provided by
MEMRI JTTM

👁 139   edited 5:38 AM

On December 7, 2018, the channel gave the same Telegram number and username, WhatsApp number, and Bitcoin address: "Our winter clothing distribution we our handing out clothes packs to widows and orphans to help keep them warm this winter. Due to your donations we were able start this project. Please continue to donate. Telegram +90 537 822 0931. Whatsapp +46 73 929 9885. Telegram Admin @thefigtree. Bitcoin address:1FeocjAnzCY4BEB7pDotUAXFK9fv9774zf."

The Merciful Hands

Provided by
MEMRI JTTM

Our winter clothing distribution we our handing out clothes packs to widows and orphans to help keep them warm this winter.

Due to your donations we were able start this project

Please continue to donate
Telegram +90 537 822 0931
Whatsapp +46 73 929 9885
Telegram Admin @thefigtree
Bitcoin address:
1FeocjAnzCY4BEB7pDotUAXFK9fv97
74zf

👁 77   6:23 AM

The Coming Storm: Terrorists Using Cryptocurrency

On December 6, the channel posted an appeal for donations and gave the same Telegram, WhatsApp, and Bitcoin information.

The Merciful Hands
Assalam alakium wa rahmutalhi wa barakatu

Our winter project is still going on we have attended to quite a few a families but there are still a lot in need.

Many families have to choose between heating themselves or eating food.

Most don't have heaters and use trash cans to stay warm. This causes alot of bad things to go into the lungs.

There burning whatever they can just to survive this brutal winter and the rain has been even more severe this year then years before.

So many places have leaks in there homes and are getting flooded and none have money to fix any of this stuff.

Only with your generosity can we continue to help these families. We still have a long way to go as the time goes by the colder it gets.

So we are urgently appealing to you for donations to help us with the winter project.

We will be doing a clothes distribution drive soon to widow and orphans with winter clothing in Sha Allah.

But we are in need of funds in order to keep our projects going. No money no possible way to carry out our project.

So we appeal to you today to donate whatever you can where it be a euro a dollar 100 dollars whatever you can your effort is appreciated and should never be looked down as it is not enough.

To donate
Telegram +90 537 822 0931
Whatsapp +46 73 929 9885
Telegram Admin @thefigtree
Bitcoin address: 1FeocjAnzCY4BEB7pDotUAXFK9fv9774zf

Provided by
MEMRI JTTM

© 193   edited 6:44 AM

The next day, the channel posted a photo of a man holding bills of an unknown currency.

On March 22, 2018, the channel wrote: "We as muhajireen [immigrants, i.e., foreign fighters] are trying to integrate into the society of the Syrian people by helping them with there needs." The message describes a toy drive the channel is organizing and gives the same information in German.

The Merciful Hands
Assalam Alakium wa rahmutalhi wa barakatu

Bismillah hirahmanirahrm wa sallutu salam alla rasullua ama bad

We as muhajireen are trying to integrate into the society of the syrian people by helping them with there needs.

So we are starting a toy drive to build a bridge between the ansar and muhajireeen so please to donate to our project to help these kids smile even its for one day after 7 years of war

So if you feel this is worthy to donate to please help us to help them.

Jazakallah Khair
_____

German (Almani)
Assalam Alakium wa rahmutalhi wa barakatu

Bismillah Hirahmanirahrm wa Sallutu Salam alla Rasullua ama bad

Wir als Muhajireen versuchen, uns in die Gesellschaft der Syrer zu integrieren, indem wir ihnen bei ihren Bedürfnissen helfen.

Also beginnen wir eine Spielzeug Verteilung, um eine Brücke zwischen den Ansar und Muhajireen zu bauen. Also bitte spendet für unser Projekt, damit diesen Kindern geholfen wird, um ihnen wenigstens für einen Tag nach 7 Jahren Krieg ein Lächeln zu schenken.

Bitte helft uns, ihnen zu helfen.

Jazakallah Khair   354 edited 5:21 PM

On March 25, the channel wrote: "A brother here has recently lost his phone and doesnt have any money to purchase one... There is also a second brother his phone got broken and not able to be fixed and just like the first brother he is in need of a phone. This brother is is injured so he runs a dawa program with over 2k followers and without the work is becoming difficult. right now these issues

are very important for the brothers to stay in contact with the field managers and there families... A brother who was working doing media for the sake of Allah had his laptop, mouse, external hard drive, and usb all stolen. So now we are trying to help these brothers get back on there feet since they dont have incomes. So it has incumbent on us to help them. so whatever donation you would like to make is appreciate from 5 dollars to 200 dollars Allah knows best your intention so help the brothers just for the sake of Allah. feel free to give a phone or laptop so we donate. to help the brothers or if you like you can make a donation and we can purchase the items ourselves and give to the brothers."

The Merciful Hands
Assalam alakium wa rahmutalhi wa barakatu

A brother here has recently lost his phone and doesnt have any money to purchase one
So if any brothers can donate or get a phone or it would be appreciated

2. There is also a second brother his phone got broken and not able to be fixed and just like the first brother he is in need of a phone. This brother is is injured so he runs a dawa program with over 2k followers and without the work is becoming difficult.

right now these issues are very important for the brothers to stay in contact with the field managers and there families.

3. A brother who was working doing media for the sake of Allah had his laptop, mouse, external hard drive, and usb all stolen.

So now we are trying to help these brothers get back on there feet since they dont have incomes. So it has incumbent on us to help them.

so whatever donation you would like to make is appreciate from 5 dollars to 200 dollars Allah knows best your intention so help the brothers just for the sake of Allah.

feel free to give a phone or laptop so we donate. to help the brothers or if you like you can make a donation and we can purchase the items ourselves and give to the brothers.

jazakallah khair!!!
Brothers and Sisters.

May Allah reward all you immensly for your help to these brothers.
344   4:38 PM

March 26

The Coming Storm: Terrorists Using Cryptocurrency

An April 2 post describes a "muhajir" who needs supplies for his baby, and another who has not received his salary for two months.

The Merciful Hands
Assalam alakium

There is a muhajir brother with a new baby and has not received salary for the last 3 months things are very tight for him and he needs help. The brother can not afford nappies and baby milk.

Also

another muhajir brother has not received any wages for the last two months and doesnt know where he will get money or food boxes.

if we all come together we can help to make a difference in these families lives.

jazakallah khair

Arabic (عربي)

السلام عليكم ورحمة الله وبركاته

هناك اخ مهاجر متزوج لديه طفل جديد ولم يقبض اي منحة منذ ثلاثة اشهر فوضعه صعب حاليا ما يملك ثمن الحليب والحفوضات لطفله

وهناك اخ آخر مهاجر لم يقبض شي منذ شهرين ولا يعرف من اين سيأتي بمال او ماعونة

اذا اشتركنا ما بعضنا البعض بمكنا ان نساعد هذه العوائل في تغير حالهم للاحسن

وجزاكم الله خيرا

An April 3 post read: "Please message me on telegram or whatsapp to help give aid to those in need. +90 537 822 0931."

The Merciful Hands

السلام عليكم ورحمة الله وبركاته
Please message me on telegram or whatsapp to help give aid to those in need.
+90 537 822 0931

تواصلوا معي هنا على تلقرام او على واتس

٠٠٩٠٥٣٧٨٢٢٠٩٣١

On April 5, the channel posted a photo of someone holding two $50 bills and a piece of paper that read: "Mercyful Hand with Abu Ahmed Foundation." Text accompanying the image read: "Merciful hands, Abu Ahmad foundation. And your support jazakallah khair [may Allah reward you]." The Abu Ahmed Foundation is an Indonesian fundraising group that raises money to buy materials for jihadis using Facebook, Twitter, WhatsApp, Instagram, and Telegram.[178]

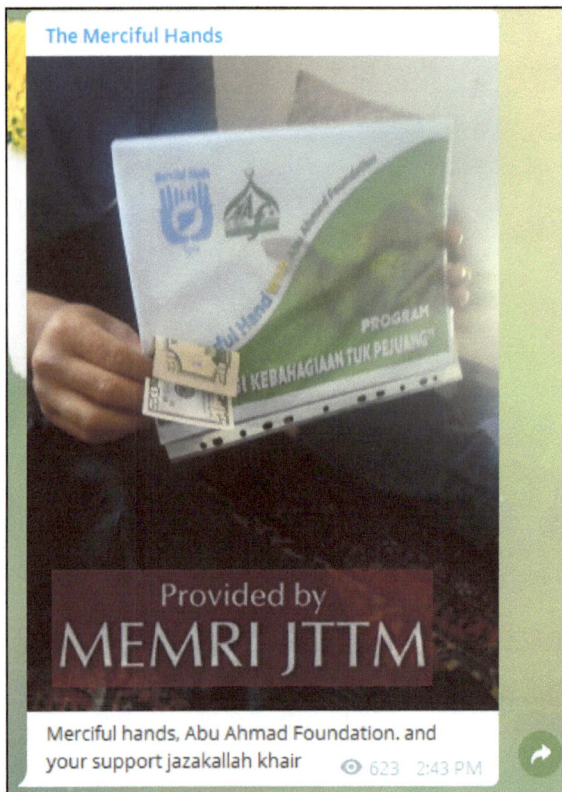

The Merciful Hands

Provided by
MEMRI JTTM

Our new logo, The merciful hands and the Abu Ahmad Foundation

Alhamdiallah another family we were able to help today   285   edited 4:08 PM

The same day, the channel wrote: "The Merciful Hands, our charity which is bringing aid to the brothers on the fronts and now we are working in conjunction with The Abu Ahmad Foundation to bring aid to people of syria."

The Merciful Hands

Provided by
MEMRI JTTM

Merciful hands, Abu Ahmad Foundation. and your support jazakallah khair   623   2:43 PM

On April 6, the channel posted a photo of a person holding two $20 bills kneeling in front of a box that read: "Merciful Hand with Abu Ahmed Foundation." Text accompanying the photo read: "Our new logo, The merciful hands and the Abu Ahmad Foundation. Alhamdiallah another family we were able to help today."

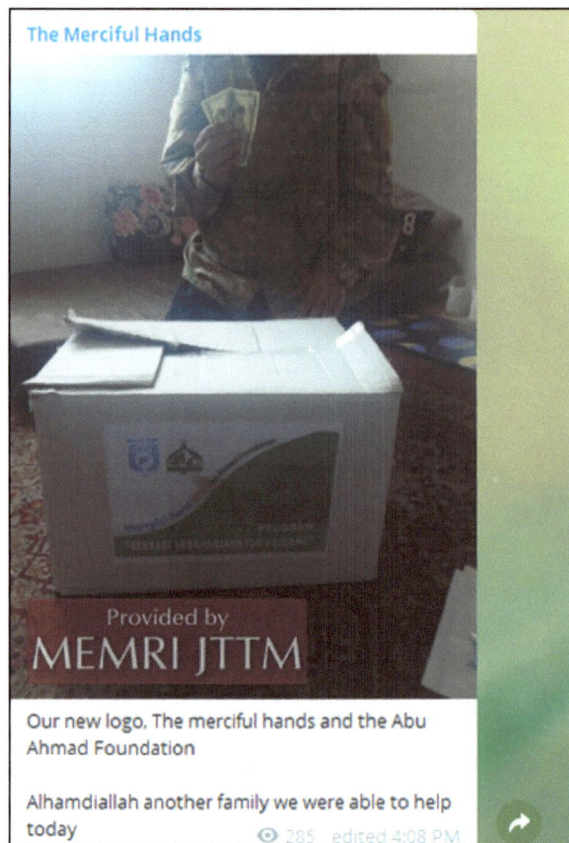

April 6

The Merciful Hands
The Merciful Hands, our charity which is bringing aid to the brothers on the fronts and now we are working in conjunction with The Abu Ahmad Foundation to bring aid to people of syria.

May Allah keep us firm and aid us amim

@aafmediacentre

Provided by
MEMRI JTTM

264   edited 4:02 PM

Also on April 6, the channel wrote: "A brother recently married decided its time to ribaat [guard duty] snd they told him if he does not have his own gear he is not allowed to come. so this is a plea to help this brother with then

The Coming Storm: Terrorists Using Cryptocurrency

equipment he needs so he can teturn He was once in a jammat [congregation] they provided everything. Now the brother wants ro return to rhe field but the jammat said there is no extra you must come equipped so the brother cried yesterday cause all he has nothing to go battlefield with what they told him broke his heart... the brother needs a baruda jaba boots snd clothing... equipping a brother in the cause of Allah is like fighting yourself dony de received shaytan." The post included the same message in German.

The Merciful Hands
assalam alakiium wa rahmutalhi wa barakatu

Urgernt 2nd appeal

A brother recently married decided its time to ribaat snd they told him if he does not have his own gear he is not allowed to come.

so this is a plea to help this brother with then equipment he needs so he can teturn He was once in a jammat they provided everything. Now the brother wants ro return to rhe field but the jammat said there is no extra you must come equipped so the brother cried yesterday cause all he has nothing to go battlefield with what they told him broke his heart.

May Allah accept from you to help equip this brother for the sake of Allah

allahumustan

the brother needs a baruda jaba boots snd clothing and money for his family as he is poor with no help rom anywhere so we as muslim ummah muslim must wake up and help with whatever we have be it physically or with our wealth we need much to help this brother

equipping a brother in the cause of Allah is like fighting yourself dony de received shaytan.

lets help this brother as much as possible

An April 9 post read: "We offer our channel in 4 different languages German, English, Turkish, Arabic."

The Merciful Hands
We offer our channel in 4 different languages German, English, Turkish, Arabic   330   10:38 AM

A post from April 27 showed a photo of a person holding a $50 bill and read: "Jazakallah khair for your donations we were able to help this poor family from another country."

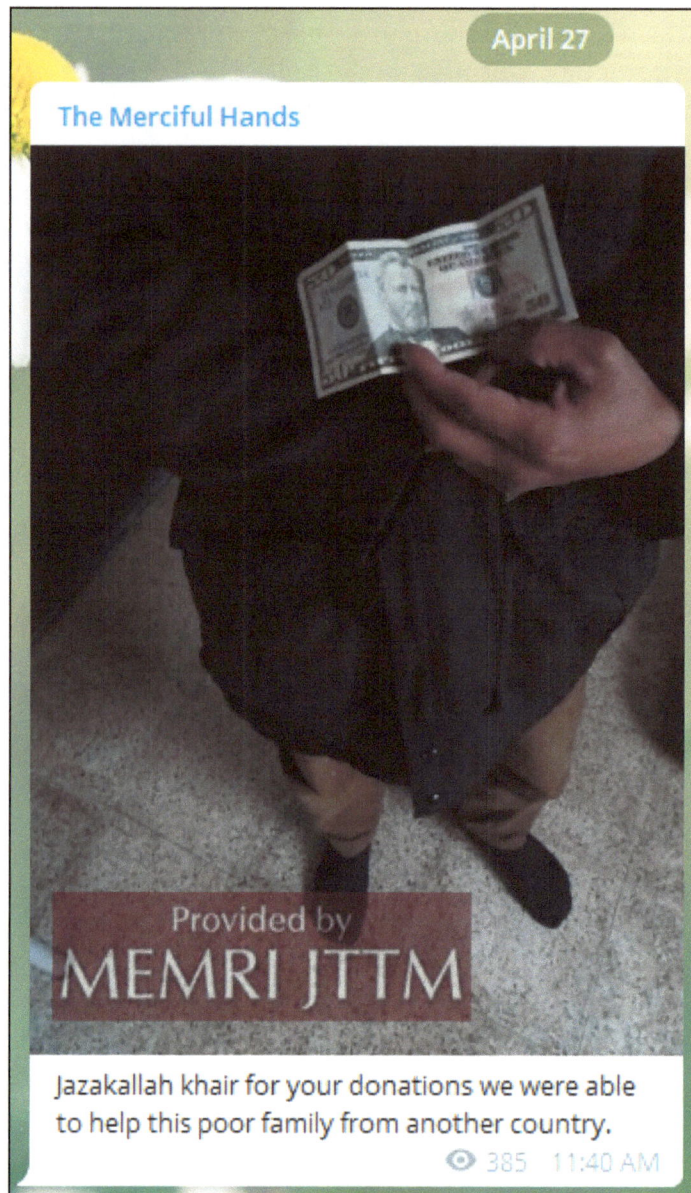

A May 6 post read: "We ran into this brother we know personally who works on the front line, but doesn't receive any payment, so he had no money and couldn't afford to buy some nessecities [sic]. So we gave him some cash and bought his food for him and his family."

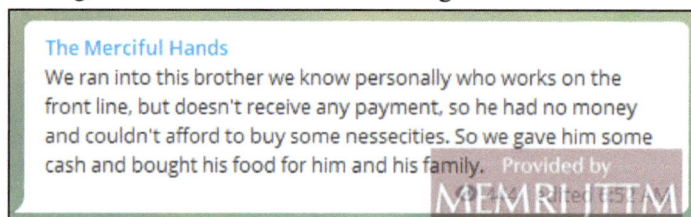

The Coming Storm: Terrorists Using Cryptocurrency

"Free A Sister" From Syrian Government Prison –Instructs Followers On Sending Bitcoin To Syria Via Paypal, Paxful, Blockchain

The Setmefreefromprison Telegram channel, created March 24, 2019 and which as of this writing has 216 members, states that it is Syria-based and that it aims to raise funds to secure the release of Sunni Muslim women imprisoned by the regime of Syrian President Bashar Al-Assad: "We humbly request you to help us so together we can set our sisters free from regime prison." The channel instructs followers, including with posts forwarded from the jihadi Telegram channel Leadmetoparadise4, on using Paypal, Blockchain, and Paxful to donate in bitcoin to the group. The channel also answers questions and explains security procedures for donating.[179]

Setmefreefromprison
Assalamualikum ,

For brother's & sisters who are asking for more info:

We are currently working for sisters who are in prison of Syrian regime.
They are in prison because they are Sunni muslimah.
Assad is just crazy & to torture or for Money capturing our sisters.
Life in prison is not like outside & for sisters it's more painful where they are literally abused in prison 😭 😭 😭

So what we do is . we try to negotiate with regime police to set free our sisters.
Sometimes to release a sister we have to negotiate with regime police & depending on case of sisters they ask 1k$ or 2K$ or 3/4/6/7/8/9/10k dollars.
Just for one sister.

But if we collect together then it's not burden on us or a single person.

So if you can donate 20$. 50$. 100$ or 1k$ or more ...
By collecting from everyone we may set free a sister of ours.

When we get success to release a sister we will try to make a video from that free sister and post in our channel.

Please to donate contact here:
@setmefreefromprison

Jazakallah-khaer

"Fragrance Of Martyrs" Explains How To Send Funds In Bitcoin For Purchasing Weapons

The "Fragrance of Martyrs" channel on Telegram, created March 8, 2019, has 384 followers as of this writing, and states that its purpose is "to incite the believers."[180] Shortly after its launch, the channel posted a message forwarded from another channel, "Helping Hands of the Mujahideen," that explained how to send funds in bitcoin for the purchase of "klashnikov, pistol, grenades, knives..."

عطر الشهداء - Fragrance of Martyrs
Forwarded from Helping Hands of the Mujahideen
Assalam alakium wa rahmutalhi wa barakatu

This is a our new channel for donations and dawa for jihad.

We are in a severe lack of financial support, and we need help to help the brothers continue on with there jihad

Including buying
Military clothing
Boots
Klashnikov
Pistol
Grenades
Knives
Jaba

Provided by
MEMRI JTTM

We are a movement located in idlib, Syria.

We are not connected to any group we are only connected to brothers who are fighting fi sa billlah.

To donate anonymously
We use bitcoin, but we have various way you can donate..

Telegram +31 6 87 59 76 56
Whatsapp +46 79 043 48 32
To contact us, also share our page

@helpinghandsofmujahideen   👁 329   1:20 AM

240

"Ummah Reports" Solicits Funds Via Bitcoin, Provides Bank Account To Support Family Of Slain Kashmiri Militant, Warns Against Using WhatsApp

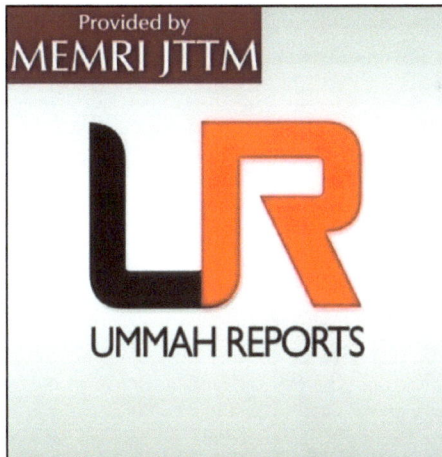

The Ummah Reports logo.

The Ummah Reports Telegram channel, created on November 21, 2018, highlights the plight of Muslims, especially in Syria, Palestine, Yemen, and Kashmir, as well the plight of the Rohingya and Uyghur Muslims. Aside from posting content illustrating these topics, the channel solicits funds via Bitcoin and by providing bank account information.

The Ummah Reports Telegram channel description reads: "Reporting the unreported. Raising the awareness of people who suffering around the world." The channel uses a Telegram bot (@ummahreports_bot) for contact.

The channel fundraises and posts Bitcoin and bank account information for a variety of causes. On December 9, 2018, the channel forwarded a message by Telegram user Sabranyanafsi (@ Nasr15), who says he is in Syria, soliciting donations via Bitcoin for families in need. The message read: "Hope this reaches you in the best of health and *iman* [faith]. I am a brother here in sham. If you know anyone that wants to help support families here *fisabillah* [for Allah's sake], it can be done via bitcoin anonymously insh'allah." Sabranyanafsi provided the Bitcoin wallet address 1D2tUhZJ VMgiB7dVRgYtA1s7ZsLAuwqH L7. As of this writing, the address shows zero transactions.[181]

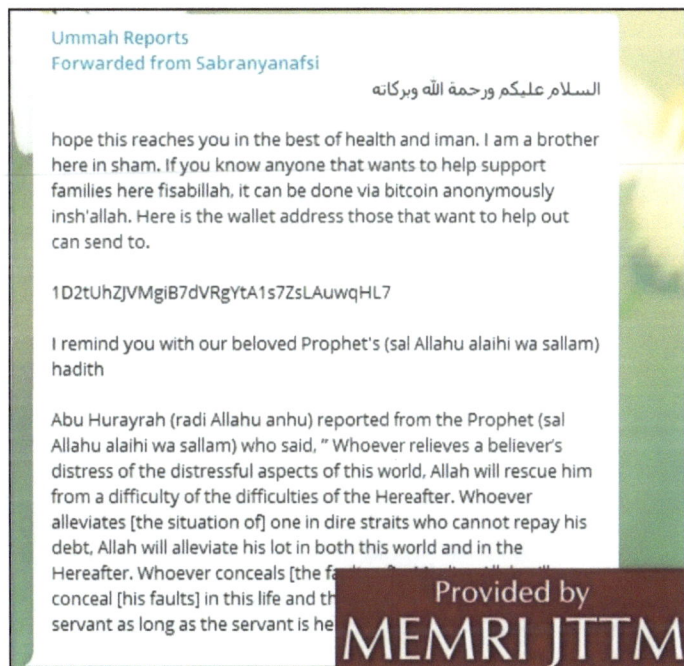

On December 13, 2018, Ummah Reports published a poster asking donations to be sent to the family of slain Kashmiri militant Zahid Bhai aka Zahid Mir, who was killed in October 2017. The poster read: "Our Mujahid brother Zahid who got his martyrdom and was the only lone son among his four sisters. His two sisters have been married and we have still got two [remaining] sisters. Alhamdulilah, here is our chance to help the sisters of Shaheed

Zahid Bhai: Let's make some Sadaqah [charity] through online." The poster included the following bank account information:

Bank Account Name: Rabia Jan T/H Father
Bank Account Number: 12440440800004155
Branch J&K Bank HR Shopian
IFSC Code: JAKAOHRSHOP
The post included a link to the Telegram bot "Charitybot," which is affiliated with Sadaqa,[182] a charity group that fundraises for the families of slain Kashmiri militants.

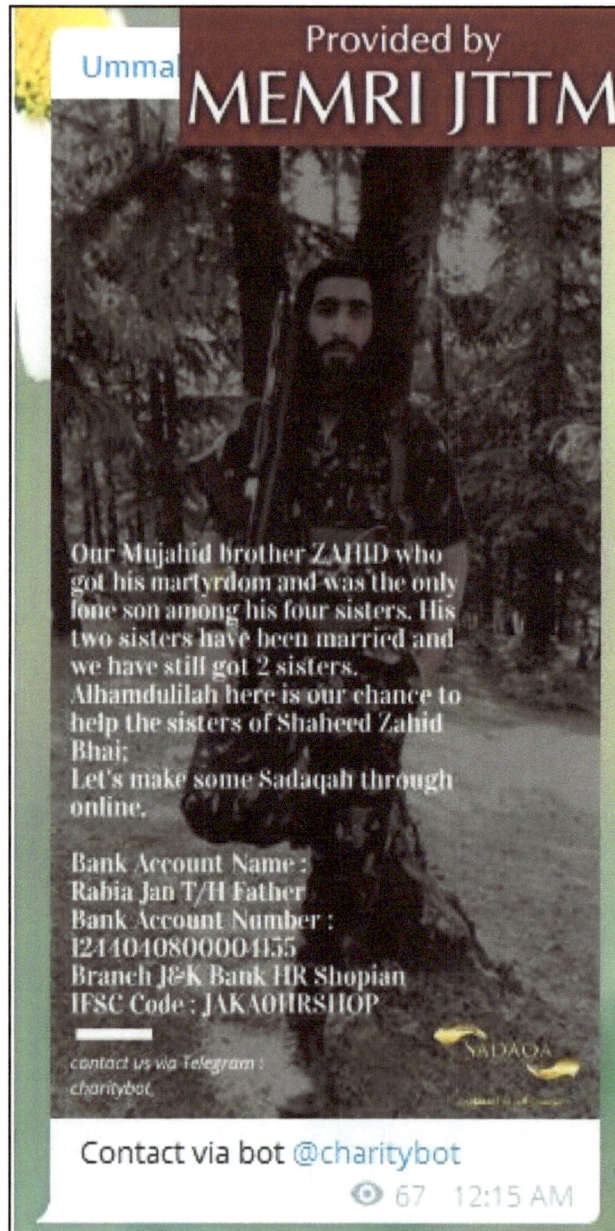

The poster with the "charity" bot contact at the bottom.

The Coming Storm: Terrorists Using Cryptocurrency

The Telegram account of the charity bot (@Charitybot), which uses the name "A Charity Foundation For The Needy."

The Ummah Reports posts other content as well. For example, on December 13, 2018, the channel forwarded a message from the Ahle Wafaa Telegram channel, warning users not to use WhatsApp. The post read: "If you are using WhatsApp and you delete the msgs after chatting, there's a place where your msgs are still stored and the agencies can access them very easily. Go to File Manager, WhatsApp, Backups and Databases, and Delete both backups and databases. This will somehow help a bit. Avoid using WhatsApp in the first place."

"Shaam Witness" Is Pro-Hay'at Tahrir Al-Sham, Provides News Updates On Fighting In Syria And Islamic Content

The Shaam Witness Telegram channel, launched March 13, 2018, currently has 1,457 followers and opened on March 13, 2018. The channel provides news updates, primarily regarding the current regime offensive in Idlib and Northern Hama in Syria, as well as Islamic content such as prayers and short religious texts. The channel is anti-ISIS and pro-Hay'at Tahrir Al-Sham, and posts HTS fighters' death announcements and celebrates their victories over Syrian regime forces. It also often highlights HTS activities against ISIS in Idlib. It disseminates and promotes fundraising campaigns from other channels, most of which claim to be charities; some of them are aimed at supporting fighters.

On June 12, 2019, Shaam Witness forwarded a post from the Bitcoin Transfer Telegram channel, that explained, in English, Turkish, Russian, and Arabic, about the company's services and promised "total anonymous and safe" transfer of funds. It also gave several Telegram accounts to contact.

The Coming Storm: Terrorists Using Cryptocurrency

BITCOIN TRANSFER

BITCOIN TRANSFER company is opening and propose to you to receive your money from all over the world, in a total anonymous and safe way You want to receive money from Europe, Saudi, Asia, Africa and America and you don't want to use unsafe transfer means for the one sending you the money? BITCOIN TRANSFER is the solution
For more details contact us privately

BITCOIN TRANSFER şirketi, isimsiz ve güvenli bir şekilde dünya çapında bir para kabul hizmeti açmakta ve sunmaktadır. Avrupa, Arabistan, Asya, Afrika ve Ameri-ka'dan para almak istiyorsunuz ve sizi gönderen kişiyi tehlikeye atabilecek yollar kullanmak istemiyorsunuz, bu yüzden BITCOIN TRANSFER sizin için bir çözüm. Daha fazla ayrıntı için lütfen bizimle özel olarak iletişime geçin

Компания BITCOIN TRANSFER сообщает о своём открытии и предлагает вам услуги по переводу денежных средств совершенно анонимным безопасным способом с любой точки мира.
Хотите получить деньги из Европы, Саудовской Аравии, Азии, Африки и Америки и не хотите использовать небезопасные средства перевода для того, кто отправляет вам деньги?
BITCOIN TRANSFER - спешит вам на помощь!
Для получения более полной информации свяжитесь с нами лично.

شركة BITCOIN TRANSFER تقدم خدمة استقبال الأموال من جميع أنحاء العالم بطريقة سرية وآمنة. تريد تحول أموال من أوروبا أو الجزيرة العربية أو آسيا أو أفريقيا أو أمريكا ولا ترغب في المرور بطرق قد تعرض الشخص الذي يرسلك للخطر ، فإن BITCOIN TRANSFER هو الحل لك. للمزيد من التفاصيل يرجى الاتصال بنا.

@Tasmimdesigncontact

@BitcoinTransferContact

https://telegram.me/BTCTransfer

Steven Stalinsky is Executive Director of MEMRI; R. Sosnow is Lead Editor at MEMRI; M. Khayat, M. Al-Hadj, N. Mozes, A. Smith, R. Green, S. Benjamin, C. Caruso, E. Barret, and J. Goldberg are Research Fellows at MEMRI.

Endnotes

1   See MEMRI CJL report Syrian Electronic Army Soliciting Bitcoin Donations For Linux Distribution, October 24, 2014.

2   See MEMRI JTTM report The 'Dark Web' And Jihad: A Preliminary Review Of Jihadis' Perspective On The Underside Of The World Wide Web, May 21, 2014.

3   Rsis.edu.sg/rsis-publication/icpvtr/co18075-cryptocurrencies-potential-for-terror-financing/#. W7zOI2hKiUk, April 30, 2018.

4   Justice.gov/opa/file/477371/download, accessed June 11, 2015.

5   Justice.gov/usao-edny/pr/long-island-woman-indicted-bank-fraud-and-money-laundering-support-terrorists, December 14, 2017.

6   See MEMRI JTTM report Article On Bitcoin In Pro-Al-Qaeda Magazine 'Al-Haqiqa': 'We See Lots of Potential For The Use Of Cryptocurrencies For Our Purposes, But We Also See A Lot Of Obstacles', February 5, 2018.

7   See MEMRI JTTM report Pro-ISIS Website Promoted On Telegram, Solicits Donations ⊠In Bitcoin, December 4, 2017.

8   See MEMRI JTTM report Charity Group Providing Aid To Syrians Shares Its Bitcoin ⊠Address, January 24, 2019.

9   Source: Telegram/alraia_ps.

10   Cash.app, accessed February 26, 2019.

11   See MEMRI CJL report Pro-Jihad NYC Imam Who Encouraged Hatred Of Non-Muslims Discusses Cryptocurrency On Salafi Telegram Channel, Raises Money On Patreon, May 30, 2019.

12   See MEMRI JTTM reports Jihadi Social Media – Account Review (JSM-AR): Vancouver Man Posts Sermons By Al-'Awlaki, Ahmad Musa Jibril, Abdullah Faisal, Others On Facebook, December 4, 2017, and Jihadi Social Media – Account Review (JSM-AR) – Part II: Vancouver Man On Facebook Posts Al-'Awlaki Lectures, Videos By Pro-ISIS Clerics Ahmad Musa Jibril And Sheikh Faisal; News On Orlando Nightclub Attack; Neo-Nazi Content, Photos Of Guns, May 17, 2018.

13   See MEMRI JTTM report Jihadi Social Media – Account Review (JSM-AR): Man On Facebook, Previously Thought To Be In Washington, D.C.-Area, Supports ISIS, Anwar Al-'Awlaki, Abdullah Al-Faisal, Ahmad Musa Jibril; Advises On Police Surveillance, Asks About Bitcoin, August 31, 2018.

14   Seattlepi.com/local/article/Fake-veteran-gets-5-month-sentence-1250322.php, September 21, 2007.

15   See *MEMRI JTTM report* Jihadi Social Media – Account Review (JSM-AR): UK-Based Disciple Of Anjem Choudary Active On Facebook, Reportedly Arrested; Facebook Friends Include Numerous Prominent Western Jihadis Involved In Terrorist Attacks, *September 20, 2018.*

The Coming Storm: Terrorists Using Cryptocurrency

16   Nytimes.com/2015/06/28/world/americas/isis-online-recruiting-american.html, June 27, 2015.

17   See MEMRI JTTM reports British Islamist Organization Supports ISIS On Social Media, November 24, 2014 and One Year Later, Extremist Islamist Pro-ISIS 'Invite to Islam' Group On Social Media Continues To Radicalize, Convert Young People, December 21, 2015.

18   Popper, Nathaniel. "Virtual Currency Offerings May Hit a New Peak with Telegram Coin Sale," *The New York Times*, March 4, 2018.

19   T.me/Eb_aa_Agency12/10146.

20   T.me/Eb_aa_Agency12/10198.

21   Engadget.com, February 5, 2018.

22   Popper, Nathaniel. "Virtual Currency Offerings May Hit a New Peak with Telegram Coin Sale," The New York Times, March 4, 2018.

23   Russell, Jon. "Telegram has raised an initial $850M for its billion-Dollar ICO," TechCrunch, February 17, 2018, www.techcrunch.com/2018/02/16/telegram-ico-850-million/.

24   Fiabvi.vg/Documents/Principal-Legislation.

25   State.gov/j/inl/rls/nrcrpt/2016/vol2/253386.htm.

26   See Pub. L. 104-132, § 302, 110 Stat. 1214, 1248, 1250. *See also* section 219 of the Immigration and Nationality Act (8 U.S.C. § 1189), justice.gov/usam/criminal-resource-manual-16-providing-material-support-designated-terrorist-organizations.

27   Treasury.gov/resource-center/faqs/Sanctions/Pages/faq_other.aspx#cyber.

28   Executive Order 13694.

29   Un.org/sc/ctc/focus-areas/financing-of-terrorism/.

30   See MEMRI JTTM report Hay'at Tahrir Al-Sham (HTS) Media Arm Publishes Article Describing Bitcoin As 'Future Currency Of Economy', April 18, 2019.

31   See MEMRI JTTM report Russian-Speaking Journalist Provides Coverage Of Hay'at Tahrir Al-Sham (HTS) In Syria; Raises Funds In Bitcoin On Telegram, April 29, 2019; MEMRI CJL report Russian Jihadi On Telegram Recommends Encrypted Private Messaging App Wickr Me, November 21, 2018.

32   See also MEMRI JTTM report Syria-Based Hay'at Tahrir Al-Sham (HTS) Jurist Abu Al-Fath Al-Farghali Declares Bitcoin Permissible For Giving To Charity, July 15, 2019.

33   Sadaqabmnor4ufnj.onion, accessed November 7, 2018.

34   Archive.is/WAP3f, accessed November 7, 2018.

35   The Bitcoin address is 14gymFijxkFzbxbacbP9ioGndsqHRuJJTc.

36   The Monero address is 42eTgdYrtCvBWxHhnLxZY3MKwouijLwaCLMMyD8YhwDvfptQkLCSb-JXYZCcVk6EB3gNHVvXNuDgqq24dk9mxqH9mxqHoBJY8L7PV.

37   See MEMRI TV Clip No. 6929, Syrian Jihadi Cryptocurrency Crowdfunding Platform Releases Video of Militant Leader Encouraging Supporters to Donate Through Bitcoin, January 3, 2018.

38 See MEMRI JTTM report Profile Of A Group Fundraising Online In English – In Bitcoin – For The Jihad In Syria, November 9, 2017.

39 Taqī ad-Dīn Ahmad ibn Taymiyyah, Al-Ikhtiyaaraat, p. 530.

40 T.me/alsadaqah2, November 9, 2017.

41 T.me/sharegroupAlsadaqah, November 9, 2017.

42 Bitcoin wallet address: 15K9Zj1AU2hjT3ebZMtWqDsMv3fFxTNwpf

43 Telegram groups such as: T.me/ummahcgr, T.me/media islamic group, T.me/HTSmedia.

44 Telegram account Musa Muhammad (@abumusa246).

45 Telegram account Javid Seyad Ahmed (@IbnulHindi).

46 See MEMRI JTTM report Jihadis In Syria Solicit Funds On Telegram, Instagram In Bitcoin To Reinforce Military Facilities, December 4, 2017.

47 See MEMRI JTTM report Pakistani Hay'at Tahrir Al-Sham (HTS) Fighter Solicits Donations For 'Poor Syrian Mujahideen,' Offers Assistance Via WhatsApp, Telegram In Immigrating To Syria, November 21, 2017.

48 See MEMRI JTTM report Profile Of A Group Fundraising Online In English – In Bitcoin – For The Jihad In Syria, November 9, 2017.

49 See MEMRI JTTM report Foreign Fighter In Syria Reports Successful Social Media Fundraising Of Bitcoin To Improve Conditions In The Trenches, December 19, 2017.

50 Facebook.com/alsadaqahofficial, December 17, 2017.

51 Twitter.com/alsadaqah1, December 17, 2017.

52 Instagram.com/alsadaqah1, December 17, 2017.

53 T.me/AlSadaqah1, December 17, 2017.

54 See MEMRI JTTM report Pakistani Engineer On Twitter With HTS-Affiliated Bitcoin Fundraisers In Syria: Give 800M Accurate Semi-Auto .338 Sniper Rifles With Thermal Sights, All Weather Uniform, December 21, 2017.

55 Telegram.me/AlSadaqah1, January 16, 2018.

56 See MEMRI JTTM report Al-Sadaqah Organization Touts 'New And Completely Anonymous' Way For Donating To It In Bitcoin, Instructs Donors To Send Vouchers Via Telegram's Secret Chat, January 18. 2018.

57 See MEMRI JTTM reports Jihadis In Syria Solicit Funds On Telegram, Instagram In Bitcoin To Reinforce Military Facilities, December 4, 2017; Profile Of A Group Fundraising Online In English – In Bitcoin – For The Jihad In Syria, November 9, 2017.

58 Telegram.me/AlSadaqah1, January 19, 2018.

59 See MEMRI JTTM report Al-Sadaqah Organization Asks People To Use Bitcoin ATMs To Send Money Anonymously To Mujahideen In Syria, January 22, 2018.

The Coming Storm: Terrorists Using Cryptocurrency

60 See MEMRI JTTM report Al-Sadaqah Organization Launches Website, Promotes Use Of Bitcoin, Other Cryptocurrencies To 'Sponsor' Mujahideen, Purchase Weapons In Syria, February 15, 2018.

61 See MEMRI JTTM report Al-Sadaqah Organization Touts 'New And Completely Anonymous' Way For Donating To It In Bitcoin, Instructs Donors To Send Vouchers Via Telegram's Secret Chat, January 18, 2018.

62 Twitter.com/AlSadaqah1, February 9, 2018.

63 Telegram.me/alsadaqah4, February 10, 2018.

64 Mozello.com.

65 1BQAPyku1ZibWGAgd8QePpW1vAKHowqLez.

66 15K9Zj1AU2hjT3ebZMtWqDsMv3fFxTNwpf.

67 Blockchain.info/address/1BQAPyku1ZibWGAgd8QePpW1vAKHowqLez, February 15, 2018.

68 @Alsadaqahsyria.

69 Al-sadaqah.mozello.com/information, February 15, 2018.

70 Al-sadaqah.mozello.com/fa, February 15, 2018.

71 Al-sadaqah.mozello.com/fa, February 15, 2018.

72 Al-sadaqah.mozello.com/page-2/wage, February 15, 2018.

73 Al-sadaqah.mozello.com/page-2/wage, February 15, 2018.

74 Al-sadaqah.mozello.com/page-2/equipment, February 15, 2018.

75 Al-sadaqah.mozello.com/page-2/equipment, February 15, 2018.

76 See MEMRI JTTM report A Review Of Al-Sadaqah Organization's Activity On Telegram: Group Updates Its WhatsApp Number, Uses Deep Web-Based Email Service SecMail August 6, 2018.

77 Secmailw453j7piv.onion/src/login.php, accessed August 6, 2018.

78 Secmailw453j7piv.onion/src/faq.php, accessed August 6, 2018.

79 Telegram.me/alsadaqah6, accessed August 6, 2018.

80 Telegram.me/Tasmimdesign, August 11, 2018.

81 See MEMRI JTTM report New Poster Appeals For Funds For Syrian Jihad Via WhatsApp And Bitcoin, August 13, 2018.

82 Telegram.me/alsadaqah6/45, July 2, 2018.

83 See MEMRI JTTM report Jihadi Fundraising Group Al-Sadaqah Announces New Facebook Page, Gives WhatsApp Number, Bitcoin Address, October 30, 2018.

84 See JTTM report A Review Of Al-Sadaqah Organization's Activity On Telegram: Group Updates Its WhatsApp Number, Uses Deep Web-Based Email Service SecMail, August 6, 2018.

85 Nulltx.com/the-early-history-of-monero-in-500-words, November 10, 2017.

86 A ring signature is a type of digital signature that can be performed by any member of a group of users that each have keys, meaning that a message signed with a ring signature is endorsed by someone in a particular group of people.

87 Stealth addresses grant additional security to the recipient of a digital currency by requiring the sender to create a random one-time address for a given transaction.

88 Europarl.europa.eu/RegData/etudes/STUD/2018/604970/IPOL_STU(2018)604970_EN.pdf, accessed October 28, 2018.

89 Getmonero.org, accessed October 29, 2018.

90 Rand.org/blog/2017/04/are-terrorists-using-cryptocurrencies.html, April 21, 2017.

91 Bitdegree.org/tutorials/monero, August 9, 2018.

92 Src.getmonero.org/legal, June 22, 2017.

93 See MEMRI JTTM report American Journalist Associated With Jabhat Al-Nusra Survives Drone Strike, June 25, 2016.

94 See JTTM report Media Outlet That Interviews Jihadi Militants & Figures In Syria Solicits Funds Via Bitcoin, Monero Cryptocurrencies, April 13, 2018.

95 Washingtonpost.com/local/public-safety/judge-allows-journalist-to-challenge-claimed-inclusion-on-us-drone-kill-list/2018/06/13/956fe70c-6f5d-11e8-afd5-778aca903bbe_story.html?utm_term=.65c422bab830, June 13, 2018.

96 See JTTM report Indonesian Charity In Syria Works With Anti-Assad Group Known To Use Cryptocurrencies To Raise Funds For Mujahideen, April 16, 2018.

97 See MEMRI JTTM report Syria-Based Twitter Account Promotes Jihadi Groups, Raises Money Via Bitcoin, Monero, Links To On The Ground News (OGN) Patreon Page, Uses Curious Cat Q&A Platform, May 22, 2019.

98 See JTTM report For The Jihad In Syria: New Cryptocurrency Crowdfunding Project On The 'Dark Web', August 30, 2018.

99 The Monero wallet address is: 8BzWKL6nXzeMjfLotpF4QXJV7pM5e2esRCYR7NQ3zXbRLKzb3Vc-c24ZCKcBpwu35HPU6nNr5JYfeTNP8MSsFLS599CTRm4s

100 See JTTM report Al-Sadaqah Organization Launches Website, Promotes ⊠Use Of Bitcoin, Other Cryptocurrencies To 'Sponsor' Mujahideen, ⊠Purchase Weapons In Syria, February 15, 2018.

101 See CJL report Anti-Assad Group Experiments With Privacy-Focused Cryptocurrency Monero To Fund Jihad, April 5, 2018.

102 See JTTM report Jihadi Social Media – Account Review (JSM-AR): NYC Man On Facebook Praises NYC Bomber, Hay'at Tahrir Al-Sham (HTS) Fundraiser, February 27, 2018.

103 Telegram.me/EHF_AR, December 11, 2018.

104 Investopedia.com/tech/what-zcash, May 14, 2018.

105 Z.cash/blog/funding, February 1, 2016.

The Coming Storm: Terrorists Using Cryptocurrency

106 See MEMRI JTTM report Pro-ISIS Electronic Horizons Foundation (EHF) Warns Against Using Unsecure Online Payment Services, Says It Will Publish Future Articles On 'Reliable' Cryptocurrencies, July 12, 2019.

107 See MEMRI JTTM report Malhama Tactical – A Jihadi Private Military Contractor Operating In Syria, Comprising Former Russian Soldiers – Trains Fighters From Al-Qaeda-Affiliated HTS, TIP, Fundraises Using Bitcoin And Selling Merchandise Online, Is Active On Social Media, October 19, 2018.

108 See MEMRI JTTM report Jihadi Military Contractor Malhama Tactical Releases Video Highlighting Trainers' Military Expertise, Asks Supporters To Donate Using Bitcoin, February 1, 2019.

109 Kipascimento.com/tr-TR/cimento/, accessed October 9, 2018.

110 MEMRI has published numerous reports on Al-Sadaqah. See for example MEMRI JTTM report A Review Of Al-Sadaqah Organization's Activity On Telegram: Group Updates Its WhatsApp Number, Uses Deep Web-Based Email Service SecMail, August 6, 2018.

111 See MEMRI JTTM report Charity Group On Telegram Solicits Money Via Bitcoin, Supports Syrian Fighters, Wives Of 'Martyrs', August 2, 2018.

112 Telegram.me/AL_ikhwa_official, August 2, 2018.

113 Blockchain.com/btc/address/1DnKvXkfAKnBnp8WzfwCFSLakoYv9y6p6S, accessed August 2, 2018.

114 See MEMRI JTTM report New Fundraising Campaign On Twitter To Arm Mujahideen in Gaza, August 2, 2015.

115 See MEMRI JTTM report Salafi-Jihadis Conduct Online 'Equip Us' Campaign To Raise Funds For Jihad In Gaza, December 15, 2015

116 The Arabic term used here, *ghurba* ("alienation," "estrangement"), is used in Salafi-jihadi discourse to convey that the true believers are a small, isolated group that is shunned and persecuted by others.

117 Islamion.com, July 10, 2015.

118 Telegram.me/jahezona02, December 2, 2015.

119 Telegram.me/jahezona02, December 2, 2015.

120 Twitter.com/jahezona_014, November 24, 2015.

121 Twitter.com/jahezona_014/status/671478529788350465.

122 Telegram.me/jahezona02, November 29, 2015.

123 Telegram.me/jahezona02, November 26, 2015.

124 See MEMRI JTTM report Fundraising Campaign To Arm Jihadis In Gaza Solicits Donations Via Bitcoin, July 7, 2016.

125 See MEMRI JTTM report Salafi-Jihadis Conduct Online 'Equip Us' Campaign To Raise Funds For Jihad In Gaza, December 15, 2015.

126 See MEMRI JTTM report Campaign To Arm Mujahideen In Gaza Continues Soliciting Donations Via Telegram, March 28, 2017.

127   Telegram.me/jahezona_40, March 6, 2017.

128   Telegram.me/jahezona_40, March 24, 2017.

129   Telegram.me/jahezona_40, March 24, 2017.

130   Telegram.me/jahezona_40, March 24, 2017.

131   Telegram.me/jahezona_40, March 18, 2017.

132   Telegram.me/jahezona_40, June 5, 2017.

133   See MEMRI JTTM report Campaign To Arm Mujahideen In Gaza Continues To Aggressively Solicit Donations On Telegram, June 6, 2017.

134   See MEMRI JTTM report Gaza-Based Jihadi Group Relaunches 'Equip A Fighter' Fundraising Campaign On Twitter, Telegram, WhatsApp, March 10, 2017.

135   See MEMRI JTTM report Gaza-Based Jihadi Group Relaunches Fundraising Campaign On Multiple Social Media Platforms, June 6, 2018.

136   Telegram.me/oumma_ps, June 6, 2018.

137   See MEMRI JTTM report Gaza-Based Jihadi Group Relaunches Fundraising Campaign Urging Muslims To Donate Using Bitcoin During Ramadan, May 8, 2019.

138   See MEMRI JTTM report Gaza-Based Jihadi Group Extends Its Fundraising Campaign To Twitter, Instagram, LinkedIn, Calls On Followers To Promote It, May 14, 2019.

139   See MEMRI JTTM report Gaza-Based Jihadi Group Posts 'Equip A Fighter' Fundraising Video, Solicits Funds Using Bitcoin, Urges Muslims To Donate To Fight 'Usurper Jews,' May 22, 2019. See also MEMRI TV Clip No. 7263.

140   See MEMRI JTTM report Gaza-Based Jihadi Group Posts 'Equip A Fighter' Fundraising Video, Solicits Funds Using Bitcoin, Urges Muslims To Donate To Fight 'Usurper Jews,' May 22, 2019. See also MEMRI TV Clip No. 7263.

141   See MEMRI JTTM report Al-Qaeda Affiliate In Syria Launches Fundraising Campaign On Telegram And WhatsApp, May 20, 2019.

142   Theguardian.com/world/2017/jul/26/eu-court-upholds-hamas-terror-listing, July 26, 2017.

143   State.gov/documents/organization/45313.pdf, April 2005.

144   Nationalsecurity.gov.au/listedterroristorganisations/pages/hamassizzal-dinal-qassambrigades.aspx, accessed February 4, 2019.

145   Police.govt.nz/advice/personal-community/counterterrorism/designated-entities/lists-associated-with-resolution-1373, accessed February 4, 2019.

146   Theguardian.com/world/2017/jul/26/eu-court-upholds-hamas-terror-listing, July 26, 2017.

147   See MEMRI JTTM report Gaza-Based Jihadi Group Relaunches Fundraising Campaign On Multiple Social Media Platforms, June 6, 2018.

148   MEMRI JTTM report Hamas Military Wing Al-Qassam Brigades Calls On Supporters Worldwide To Send Funds Using Bitcoin, January 30, 2019

The Coming Storm: Terrorists Using Cryptocurrency

149   Bloomberg.com/news/articles/2019-01-30/hamas-calls-on-supporters-to-donate-to-group-in-bitcoin, January 30, 2019.

150   MEMRI Special Dispatch No. 7878, Gazan Academic And Journalist Hussam Al-Dajany: People Can Now Donate To Hamas Using Bitcoin Without Fear Of Getting Caught, Only Iran Supports Us And Is Not Ashamed Of It, February 6, 2019.

151   T.me/qassambrigades/12537, February 2, 2019.

152   T.me/qabasat2018/457, Alqassam.net/arabic, accessed February 4, 2019.

153   T.me/qabasat2018, February 5, 2019.

154   T.me/qassambrigades/12669, February 10 and 17, 2019.

155   Cryptomenow.com, February 13, 2019.

156   Sputniknews.com, February 13, 2018.

157   See MEMRI JTTM report Hamas Military Wing Al-Qassam Brigades Releases Video Of How To Send Secure Donations Via Bitcoin, March 26, 2019.

158   MEMRI JTTM report Hamas Military Wing Al-Qassam Brigades Releases Video Of How To Send Secure Donations Via Bitcoin, March 26, 2019.

159   MEMRI JTTM report Syria-Based Jihadi Preacher Urges Support For Hamas Military Wing Via Bitcoin Donations, March 27, 2019.

160   See MEMRI JTTM report Hamas Military Wing Al-Qassam Brigades Announces Plan To Send Out 'Hundreds Of Millions' Of Text Messages In Multiple Languages Explaining How To Send Funds Via Bitcoin, June 3, 2019.

161   Justice.gov/usao-nj/pr/somerset-county-man-charged-attempts-provide-material-support-hamas-making-false, May 22, 2019.

162   Nj.com/news/2019/05/he-lived-in-a-wealthy-nj-suburb-and-played-in-a-high-school-band-how-did-he-end-up-accused-of-aiding-terrorists.html, May 26, 2019.

163   Justice.gov/usao-nj/pr/somerset-county-man-charged-attempts-provide-material-support-hamas-making-false, May 22, 2019.

164   Nj.com/crime/2019/05/i-want-to-bomb-trump-tower-nj-man-charged-with-supporting-terror-group-said-feds-claim.html, May 23, 2019; Justice.gov/usao-nj/pr/somerset-county-man-charged-attempts-provide-material-support-hamas-making-false, May 22, 2019.

165   Justice.gov/usao-nj/pr/somerset-county-man-charged-attempts-provide-material-support-hamas-making-false, May 22, 2019.

166   Nj.com/crime/2019/05/i-want-to-bomb-trump-tower-nj-man-charged-with-supporting-terror-group-said-feds-claim.html, May 23, 2019.

167   Justice.gov/usao-nj/pr/somerset-county-man-charged-attempts-provide-material-support-hamas-making-false, May 22, 2019.

168   See MEMRI JTTM report Pro-ISIS Channel Shares Guide To Obtaining Weapons From Dark Web, Urges Muslims To Attack In The West, April 16, 2019.

169   T.me/leadmetoparadise4, October 13, 2018.

170   Teespring.com/Tawheed2#pid=2&cid=573&sid=front, accessed December 17, 2018.

171   Gofundme.com/islamic-guidance, accessed December 18, 2018.

172   Gofundme.com/rationalbeliever, accessed December 18, 2018.

173   See MEMRI JTTM report Indonesian Jihadi Fundraising Group Abu Ahmed Foundation Gives Bank Account Numbers, Works With Other Jihadi Organizations Including Malhama Tactical And Al Ansaar On Facebook, Twitter, WhatsApp, Telegram, Instagram, October 31, 2018.

174   Youtube.com/watch?v=j8rDHAQ09BM&feature=youtu.be, October 21, 2018.

175   T.me/joinchat/AAAAAFNenUfuuVbj9Y2lrQ, March 28, 2018.

176   See MEMRI JTTM report Telegram Channel Solicits Funds Via Bitcoin, Provides Bank Account To Support Family Of Slain Kashmiri Militant, Warns Against Using WhatsApp, December 13, 2018.

177   Telegram/The Merciful Hands, December 6, 2018.

178   See MEMRI JTTM report Indonesian Jihadi Fundraising Group Abu Ahmed Foundation Gives Bank Account Numbers, Works With Other Jihadi Organizations Including Malhama Tactical And Al Ansaar On Facebook, Twitter, WhatsApp, Telegram, Instagram, October 31, 2018.

179   See MEMRI JTTM report Jihadi Telegram Channel Instructs Followers On How To Send Bitcoin To Syria Using Paypal, Paxful, Blockchain To 'Free A Sister From [Syrian Government] Prison', April 17, 2019.

180   See MEMRI JTTM report Telegram Channel Promotes Jihad And Martyrdom, Post Quotes From Anwar Al-'Awlaki, Documents Deaths Of Fighters In Syria, April 12, 2019.

181   Blockchain.com/btc/address/1D2tUhZJVMgiB7dVRgYtA1s7ZsLAuwqHL7, December 13, 2018.

182   Telegram.me/sadaqah01, December 13, 2018.

The Coming Storm: Terrorists Using Cryptocurrency